Verboten Meteorology
The Primordial Speck Theory
Second Edition

By John Billen

Table of Contents

Introduction

Are we the worse for knowledge?
JB

Meteorology is one of the younger sciences. The writings of the early philosophers surely mentioned the passing of the seasons, and storms and drought. Without a rough topographical map of the Earth, only possible after the globe had been circumnavigated by travelers, could meteorologists properly conduct the working of that science. Civilization prior to Columbus was unaware the Earth was spherical in shape.

One would have to qualify that statement with the possibility that some surely speculated upon the shape of the Earth and concluded that it was a sphere. The idea that the Earth might be spherical in shape must have occurred to people back then, the Moon is a sphere, the Sun is a sphere, why not the Earth? What else could it possibly be? The proof of the Earth's spherical nature, and its dimensions, were lacking.

Meteorology ties in with electricity, and electricity was never well understood until a long time after the globe had been circumnavigated. Lightning, electric eels, and magnetic rocks have been observed by our species for a very long time, but knowledge of electricity goes back only two hundred years or so. The atomic structure of matter as it is currently understood is also a fairly new discovery. Knowing the extremely small size of atmospheric components is of importance in understanding or predicting atmospheric events.

Aviation, as well, or at least hot air balloons, gives an investigator into meteorological phenomenon a much enhanced perspective to observe clouds and such, and humans mastering flight is also a relatively recent event. Some satellite providers have a channel with a continuous view of the Earth from space. Quite a difference between the current perspective on things and the more limited perspective of ancient times. If Confucius or Aristotle could view a satellite image of Earth now, he would be amazed by what everyday people currently find commonplace. Time marches inexorably onward, and the day will come when we are the primitives with limited perspective. Two thousand more years will elapse, and civilization will certainly look back upon the era when humanity grappled with weather modification for the first time with some amusement.

Nowadays the skies are very well monitored on a continuous basis. New technology like Doppler radar make the observation of weather phenomenon easier than ever, with meteorologists giving more advanced forecasts and timely bad weather warnings, as often as not. The amount of data that computers integrate and store in arriving at weather forecasts is now very considerable.

A meteorological discovery appears to have taken place a little more than 110 years ago that very few people seem to realize even exists. That being my only dissatisfaction with the scientific community at present, at least since the television show about the 4.4 million year old hominid fossil discovered in the highlands of Ethiopia, that is what I've felt compelled to write about. That a discovery was made, yet is not in the encyclopedia, is not supposed to happen in our world; the discoverer mentions the discovery to someone else, or obtains a patent, eventually the whole world finds out. Why that hasn't happened in this case is open to speculation. 110 years ago, one could argue that meteorology was still in its infancy.

I am no meteorologist, nor a cosmologist, and since this book presents some ideas in both of those disciplines, all I can say is there are things to be observed in those fields that are curiously missing from any other accounts in those fields. Neither of those two scientific disciplines is any thing like chemistry, where a typical scientist can be seen conducting experiments in a laboratory with ingredients in glass beakers. What do cosmologists do, but think about the cosmos? They have no specimens to work with other than astronomical observations, so anyone, basically, can read about and observe what is known in the field. Meteorology also deals with things differently than a laboratory chemist. There isn't much in the way of specimens to examine other than weather patterns, and radar displays. There are specimens, the weather as it occurs. These are transient phenomena, here one day, and gone the next. They are not tangible things that can be removed from their natural location and examined under the microscope,though a hail specimen could qualify in that regard. What hail consists of is already known,however, so studying one hail specimen isn't likely to yield much else. Weather phenomenon are things taking place involving enormous numbers of gases blanketing the entire planet. Given the size of the Earth, the totality of water vapor being created is huge, far too much for any one location.

This doesn't appear to be what meteorologists do

Individually, these gases are invisible, but in huge numbers can accumulate into weather systems as large as hurricanes. The science of meteorology is the application of the laws of physics and chemistry to the study of the atmosphere of our planet. Thus, anyone with an education is capable of contemplating weather phenomena, and perhaps observing something new.

The main purpose when I write about this is gaining the attention of the world about the theory I have, and gaining enough public support to persuade the scientific community to give this theory a more accurate and detailed analysis in the hopes of this theory emerging as truth. Once that is actually accomplished, if indeed the truth is determined to fit the theory,one should then be able to look up Weather Modification in just about any encyclopedia and actually find it there.

I would also like to live in a more habitable world. The two events are interconnected. Living conditions won't improve if a process exists that could create perfect growing conditions, an absence of damaging weather, and higher polar ice caps, and is not included in encyclopedia. If the process I present in this book really works, about the only way for people to find out about it, with the exception of reading this book,would be to find it in an encyclopedia. Once the placement in the encyclopedia happens, the world would start becoming a better place.

The theories brought forth here, if eventually proven valid, could present the entire population of the world with a new alternative to passively accepting whatever weather comes along, good or bad. The local inhabitants of an area changes at national boundaries, and with weather systems being a natural event that respects no boundaries, there will be sure to be some overlapping effects that would spill over to neighboring areas. Hopefully everyone who reads this will see the necessity of having all the details discussed put in the proper perspective, since any attempt to alter the weather, at least from the viewpoint of the theory put forth here, would be a local phenomenon, confined to a few hundred square miles.

A person living 300 miles from an ongoing drought

situation isn't the person ideally located to change weather conditions, should that be deemed necessary. Neither is that person impacted much by the drought 300 miles away,except perhaps by rising prices for some foods. Therefore, should weather modification be discovered to be possible, those living in specific areas would see to that area, and one would expect the rest of the world to be doing pretty much the same thing. The world would become a better place by having the processes described herein available to the public since whenever a drought situation arose some child living in the area would likely look up weather modification in the encyclopedia and find that there is indeed a remedy worked out for drought included therein.

Local residents being those impacted by any weather happening where they live, provided the encyclopedia provides information currently known, are then responsible for the area in which they live. Should a drought unfold, someone living where the drought is taking place should take the steps necessary to produce precipitation. Needless to say, too much precipitation can be even worse than drought. Stopping or preventing flooding can also be accomplished using the processes described in this book. A means of preventing flooding, if it were really possible,would be a local responsibility as well. Such a process should also be included in current encyclopedia, available to the curious.

As with any knowledge held by the human species,not everyone will know this; the grand totality of human knowledge is never held by any one individual, we all hold differing bits and pieces of the total. That is the reason for record keeping of any kind in the first place.

Since a compendium of knowledge such as an encyclopedia contains the stored wisdom of the species, whenever a question arises pertaining to a certain field, the encyclopedia is often the first place to start. The mortise and tenon arrangement used to fasten pieces of furniture together was developed once. No need to be discovered, or worked out again after that since it was thereafter a part of recorded knowledge.

It may have taken some time between the first practice of that type of carpentry and the placement of the technique in an encyclopedia, but that is normal for any new idea, method, or process. Currently, if I want to build a chair, I don't have to start from scratch and develop the means to fasten it together by myself. I can look it up in the encyclopedia, or read a book about carpentry. The same conditions apply to any field of endeavor and the stored knowledge in the field. That in all probability I would buy a chair at a furniture store that has already been manufactured by experienced carpenters at a considerable savings of time, trouble and expense is immaterial. I might want to indulge in carpentry as a hobby.

Scientists should be precise and brutally honest about what has been discovered. If the scientific community doesn't sign off on an idea, that is, if they do not follow up and confirm the validity of the idea, it has no chance. There should not be any political motivation to fail to mention things that have been discovered, after some time has passed and there are no existing patents. Scientists often work on proprietary material of some corporation that has expended time and assets in the acquisition of this or that process. I don't have a problem with scientists withholding certain information in the financial interests of the corporation they work for. A long time has passed since the first time anyone noticed anything pertaining to what this book is about. There will be no patent for a process that arguably has been around for over 110 years, if I or anyone else were to apply for it.

Not all scientific inquiry is private. Many colleges and universities have ongoing scientific investigations, along with non profit charitable institutions, and most information is shared with the public, if something new is discovered. These institutions have dedicated themselves to the pursuit of truth; I can't see why some of the scientists or future scientists in this group didn't resolve the questions raised in this book long ago.

The first chapter will start with some basic discussion about weather in general. Then the start of the theory, and a start on the entire explanation of it.Chapters two and three are the cosmological parts of the book. That cosmological discoveries could have implications in meteorology is one of the main points of this book; the dynamics of the weather is what humans seek to know. If those dynamics change when one factors in huge quantities of as yet undetected particles, we should know that also. How tiny things behave in space can be answered correctly only if we are correct in knowing whether space is a fabric of some kind, or mere emptiness.

That which can be extrapolated from what we know now, by just applying common sense, seems to have been sorely missed by what little I have read about cosmology.There is surely a wealth of material I haven't seen, but pointing out what I had observed about the universe in general seemed in keeping with my observation that dark matter and dark energy might change the dynamics of the weather. After all, a theory has all of the details that need to be brought out that pertain to the processes hypothesized by the theory.

The fourth chapter discusses the meteorological aspects of the process the author purports to solve all our problems in the environmental arena. Chapter 5 is a bit of discussion about this whole concept in general, and the impact it might have on the general public of the entire world. Then we get to some of the benefits, mostly, of this new device or devices and also the costs of letting this slip through our fingers in Chapter 6. The next chapter takes a look ahead, assuming that the theory is true, and tries to show what might follow should civilization adjust to a more modern view of weather modification.

Chapter 8 gives some anecdotal evidence, followed by what are really more like appendix chapters. Chapter 9 is a look back to how and where I realized for the first time that weather modification was probably possible and is included for the sake of completeness, and the final chapter consists of suppositions and predictions that should be included, also as a kind of reference. After all, time marches on. Anyone who proposes a new theory in the natural sciences will give details and predictions of the theory. Will the theory stand the test of time, is the question. Only by making predictions and seeing those predictions borne out can the theory remain where others have fallen.

I really and sincerely hope that enough people read

this. By enough people I mean that number necessary to push this discussion beyond this book, into the real world where certain individuals compile encyclopedia, other individuals explore the science of meteorology, and still other individuals sit on the Supreme Court, and make judgments about matters involving public safety. If it turns out that only a few dozen copies of this book are all that were needed to accomplish the feat of bringing this idea further along, so be it. It has been fun putting all these descriptions and ideas down, and I could certainly say that I got something out of the activity, even if the book never sells a great many copies. The idea, or process, comes first. That is what made writing the book seem necessary in the first place. I would be more than happy to share in the fruits of more provident weather, with no income from the sale of this book.

I have tried to be as accurate with things as I could. I am convinced there is something to all that is discussed that isn't immediately apparent. A fair number of discoveries in the natural sciences involve discoveries that are tested and true but are quite invisible to the naked eye, as in the case of the force of gravity. We know it is there, but there is no putting one's finger on it, so to speak. You can't point to it and say "this is gravity", or for that matter, any of the forces of nature. They exist, surely, and are constantly at work, but their workings are only induced after long and patient observation, and take considerable explaining.

Explaining gravity by jumping, and pointing out that gravity brought the jump to a quick end does explain gravity, in the sense that one can perceive that in that instance gravity is surely at work and the huge mass of the Earth was surely responsible for the jump ending so quickly. But that doesn't explain it in it's entirety, for we have yet to discern what the cumulative particles of the Earth are actually doing to attract things towards them.

Black holes are concluded to exist, though no one has ever visited one nor is anyone likely to. They cannot be seen. They can be inferred to exist because of the tiny blank space at the center of the Milky Way, the gravitational effects observed there,and whatever other evidence. Dark Matter and Dark Energy are even more enigmatic than black holes and the forces of nature. But, indirectly, scientists have compiled enough evidence to induce that Dark Matter and Dark Energy exist, as well.

Dark matter, dark energy, and black holes are real things, surely. Processes that occur involving various things interacting with other things are not tangible things themselves, but nevertheless these processes exist and can eventually be discerned. A weather process unknown up to this point, at least in so far as not being included in encyclopedia, can be proven to exist. Said process having been discovered, the human species should then treat it in the same manner as any new process. In this instance, "weather modification" would be the title under which it should be found.

Any new process will be treated with suspicion, and this one could prove dangerous, so let the reader decide. My credentials go no further than a two year degree in accounting, with no scientific background. Surely, more clever beings exist than myself, some having spent six years or more studying meteorology. Could anyone possibly believe that I am the first and only person to have perceived something pertaining to weather modification that no one else ever has? The 1957 novel by Ayn Rand, Atlas Shrugged,may or may not have pertained to what this presentation is about. Ayn Rand passed away in 1981. Let the reader decide if the process described here isn't sufficiently similar to what the invention in that novel consisted of.

Facing weather inevitabilities head on with as much information as possible seems to me the best approach to this problem. My writing this book amply demonstrates that adopting an ostrich type of view to the processes described herein no longer is working. More individuals besides myself will discover the same things eventually. Mankind needs this issue successfully resolved. No better time than the present to more clearly define certain atmospheric events.

The weather may be the most discussed topic on the planet. Most of that discussion is almost part of the culture of many societies when greeting another person. To say "Hello, nice day today, don't you think?" can be stated in many different languages. Most weather discussion is superficial, then. In this weather discussion we are going to go where few have gone before; to wherever the clues lead us and wherever imagination, intuition, and experience lead us from there, as deeply into weather phenomenon we can get with possibilities and theories. I hope the net result is even more weather discussion around the planet.

Chapter 1.Background and Theory

"Things in high places,
with higher purposes"
JB

A lot of this presentation will be about electrical conductivity and resistance, and how these electrical processes could have an impact on weather modification, so it would be best to mention a few things about the weather. We are going to assume throughout this book that everyone agrees that experiencing pleasant, livable weather is preferred over weather that could endanger lives. Naturally the goal of any attempt to modify the weather ought to have as its purpose making the weather less dangerous, and to make more fresh water generally available. Somewhere, almost daily around the globe, some weather extremity can be found to be happening that everyone living in the area would much rather not have happening. There are currently a number of land areas classified as deserts, and some semi-arid places as well, places that could always use more water.

The Sahara Desert is the largest, about 6 million square miles, and some islands have desert on one side, some of the smallest deserts. Every time a flood occurs, that water might have been deployed elsewhere, to any number of the dry areas of Earth, had local residents of the dry areas known that they could have as much rain as they could ever want. Yes, even the Atacama Desert on the high plateau in Chile could see changes.

Correspondingly less would occur in any one local area if the entire world is intent upon obtaining an ample supply of water. Surprisingly good results look quite inevitable should weather modification become possible. The goal would be to decentralize the moisture emanating from the oceans, and also to get precipitation to fall where it is desired. Consider the grand total of water vapor over the Earth; were that viewed as a large hose of incredible power, the idea would be to cut slits in the hose at various places so that all the water doesn't wind up in one place.

That we have a blanket of atmospheric particles covering the planet that can become way too numerous in any one location is evident. Something with which to modify atmospheric quantities seems the thing desperately needed. Reducing humidity in one location by increasing humidity in twelve other locations in the same hemisphere, for example, could not only stop the imminent flooding in the one location, it would also serve to increase precipitation in twelve other arid or semi arid regions. If such a process exists that would definitely seem the most promising alternative.

The stance of the meteorological community concerning modifying the weather is that it isn't possible. Apart from some limited success with silver iodide cloud seeding, which increases cloud yields some 30% or thereabouts, no effective means of modifying the weather exists. There is a research project involving electromagnetic pulse propagation, the High Altitude Auroral Research Program, or HAARP for short, which is expected to eventually do things with the weather, but officially, not yet.

The Soviets have a complementary version of HAARP, called Woodpecker. One can find accusations on the internet that the Russians use the equipment they have to create high pressure in the middle of the Pacific, leading to drought for the western U.S.. Chemtrails, exhaust from airplanes with special additives, has been tried, as well. That would be a silver iodide cloud seeding type of experiment. Accusations have been levied against the Canadian government on the internet for conducting such activity and not telling the public about it. There have been several articles written recently that point to the reluctance of the meteorological community to share new information in the field of meteorology. These accusations involve electromagnetic pulse technology and the ignorance of the public when it comes to knowledge of what HAARP is up to. The public position is that HAARP is not currently doing anything with the weather, but there are many voices that say otherwise. A book was written titled "You Wouldn't Want To Play This HAARP" by Nick Begich,Jr. I haven't read it, but from what I understand the book argues against the government having such a device, and points out a lot of the dangers involved.

As for what one could hope to achieve if weather modification were a reality, growing more and healthier plants and animals on a world wide basis would be one benefit. No one starving to death would be a big plus. Never seeing water shortages or famine, knowing when it will rain with 100% certainty gives one a sense of well being that would be difficult to measure in monetary terms. Putting out forest fires with the new technology would also save lives and property. Making shipping lanes safer for ocean going vessels, any watercraft on the open sea, is a possibility.

An absence of hurricanes, tornados, and floods would bring an end to the endless rebuilding of structures destroyed by weather, allowing many more buildings to stand for hundreds of years. Increased precipitation in the polar regions could help to build higher ice caps, and additionally drain the oceans of the strength to generate hurricanes. Doesn't seem there is a choice for mankind should the magic wand be found that humanity could modify the weather with. We haven't reached the end of the benefits that weather control could provide, we will get to a few more later, and it seems from this point on that doing nothing with the technology, if it existed, would be far more dangerous than embracing it.

The atmosphere on our planet contains vast numbers of individual, extremely tiny things. The astounding smallness of these free floating gases is nearly impossible to grasp. That is what the science of meteorology investigates. Additionally, astrophysicists, cosmologists and others are currently maintaining that as much as 96% of all matter and energy is dark and as yet undetected. A concerted search has been mounted for dark matter and dark energy.

This implies that not only is the atmosphere comprised of tiny gases and a little dust, it is also comprised of particles even smaller in amounts as high as up to twenty four times the mass and energy of the atmosphere itself. Meteorology can be a bit more complicated when one considers all new possibilities.

There is something to be discovered involving

electrical conductivity and resistance and weather modification. The first evidence of this appears with the experiments conducted by Nikola Tesla in 1899, in Colorado Springs, Colorado with his Tesla Coils. A biography of Tesla[1] mentions that during the time when he was at Colorado Springs experimenting with his Tesla Coils on the platform above his laboratory an intense thunderstorm with some 12,000 discharges of lightning in a two hour period occurred once, with all the lightning happening within 30 miles of Tesla's lab.

The biography mentions that thunderstorms in the mountains are a common event in that area of Colorado, not far from Pike's Peak. My thinking is that the second half of 1899 saw a higher than average precipitation amount in Colorado Springs, and surrounding areas. The presence of such a large quantity of copper coils in a high place such as the Tesla Coils in Colorado Springs was the catalyst for the storm fronts that developed, by being large enough to create a path of least resistance for air molecules, dust, static electricity and possibly dark matter and dark energy to flow more freely along.

Tesla tinkering with wireless experiments inadvertently showed to anyone who happened to be watching a discovery that should have exploded onto the newspapers after a few weeks; that a large quantity of copper strategically located could change the weather. Communications being somewhat primitive at this time, something Tesla was seeking to improve, it could have taken some time for a newsworthy item to traverse the globe, and with this discovery that apparently never happened. The electricity that Tesla pumped into the atmosphere found clouds to reside in and eventually be released as lightning.The clouds were probably there because the copper was where it was.To suppose that no one noticed that the Tesla coils were changing the weather would be a serious mistake.

The magic wand was revealed for the first time, and shown to be rather too large for any one person to wield. Mankind can build the Empire State building; the means necessary to achieve this end is within human capabilities. The objective would be to place in an elevated position a quarter to half ton of copper, using whatever labor and equipment necessary. Copper is the cheapest available metal with high conductivity. Immaterial whether trucks and forklifts or manual labor are used to transport it, a precisely calculated quantity of copper in a high location is our "magic wand". Naturally, something besides magic is at work when copper in quantities meets the atmosphere. Exactly what takes place when such a placement of copper on high occurs is the question this book attempts to more thoroughly answer.

These were the first human experiments that we know of where copper in quantity was placed on high, and remained there in place for some time. The reference to Tesla having ideas about controlling the weather could pertain to this, ideas of his that he never developed. From the accounts I read of his trip to Colorado Springs his Tesla coils probably resided on the roof of his laboratory there from early June 1899 until mid January 1900 when he returned to New York.Let me give you a few parts from the same biography of Tesla;"At dusk of that day Tesla had watched a dense mass of strongly charged clouds gathering in the west. Soon the usual violent storm broke loose which after spending much of its fury in the mountains, was driven away at great speed over the plains"[2].

Pretty Big Wand

The book goes on to say he felt he made a great discovery that day; "He summed up the implications of this discovery thus: "Impossible as it seemed, this planet, despite its vast extent, behaved like a conductor of limited dimensions. The tremendous significance of this fact in the transmission of energy by my system had already become quite clear to me. Not only was it practicable to send telegraphic messages any distance without wires, as I recognized long ago, but also to impress upon the entire globe the faint modulations of the human voice, far more still, to transmit power, in unlimited amounts to any terrestrial distance almost without loss". He felt he had discovered how to transmit power cheaply that day as well as send the human voice around the globe. While reading that particular part of the book I really expected to see something about weather modification, but it wasn't there.

One other excerpt does mention weather modification; "I am positive in my conviction that we can erect a plant of proper design in an arid region, work it according to certain observations and rules, and by its means draw from the ocean unlimited amounts of water for irrigation and power purposes. If I do not live to carry it out, somebody else will,but I feel sure that I am right". The book goes on to say that "This idea too went into his legacy of unfinished business, and to this day no one has implemented it"[2].

In New York after 1899 Tesla resumed his energy transmission experiments and it has even been speculated that one of his energy transmissions caused the 1908 Siberian fireball. As it happened, Perry was just then closing in on the North Pole, and Tesla may have tried to impress Perry as he trekked the far north.

Getting back to those two sentences about the usual violent storm, the point is, the whole time that the Tesla Coils were in Colorado Springs, from mid-June of 1899 or before to at least mid-January 1900, its likely there was not one consecutive week where no storms occurred, and were probably as frequent as twice a week, and sometimes during the warm months in the mountains during low pressure periods thunderstorms can threaten in the early to mid afternoon every day for weeks on end. All summer long, then, and into early fall, thunderstorms probably threatened nearly every day, with an occasional break when a cold front followed a storm front.

Mountainous terrain is indeed quite a different

situation than rolling hills or flat prairies. As the prevailing westerlies reach the base of a mountain, the air begins to rise, especially during the hours of most intense sunlight. As the air rises it cools, and in so doing, any water vapor in the rising air condenses into clouds. When there is lower barometric pressure and more humidity in the air, those clouds can become thunderstorms very easily and quickly, blossoming in a few short hours. Florida has frequent thunderstorms for an entirely different reason, since it is one of the flattest states, topographically. The warm waters of the Gulf of Mexico create water vapor that passes right over Florida almost continuously. The high humidity and rising air in Florida during the hours of most intense sunlight causes the water vapor to condense as well, and since there is a lot of it, thunderstorms develop quickly and intensely.

In Colorado Springs in 1899, once the seasons changed and temperatures fell, the storms would most likely have become less frequent, down to around two per week. There was the discovery, way back when, and in the Colorado Springs newspaper back then the truth could be ascertained as to the frequency of storms in the area for that six or seven month period, and so the list of evidence grows, maybe. The internet probably wouldn't have newspaper archives from that long ago. I didn't go to Colorado Springs for more evidence.

Finding the exact physical causes for what happens with a large quantity of copper on a mountaintop could take considerable doing. The assertion maintained throughout this book will be that certain observable effects can be noted; these are falling barometric pressure, cloud accumulation, and precipitation, the latter usually occurring about 72 hours after placement. Since the number of experiments I conducted were limited, there would be considerable variation at other locations around the world. One would have to begin at some specific location and perceive what effects occurred there when using varying amounts of copper.

One important thing to realize, though, is that no one place on Earth is isolated from the rest of the Earth in the atmospheric sense. The atmosphere on the planet is all interconnected. If some atmospheric components begin to move faster eastward,others fill the space that has been vacated by those. It is probable that almost any location will see observable effects from placing copper strategically, one need only ascertain the particular effects at some specific location, and decide the appropriate quantity of copper to deploy so as to yield the desired sized storm.

Not all land areas on earth are mountainous. This should not prove to be an insurmountable problem when it comes to finding a suitable location to place copper where it would be one of the highest things around, and be capable of transmitting it's electromagnetic waves a long distance. In places such as the Midwestern United States where one can find only gentle hills for miles on end,the roofs of tall buildings would probably suffice to position the copper high enough to produce the desired effects.

Solid objects would block any electromagnetic waves considerably, and decrease the effective range and strength of the waves, so there is the requirement that these copper placements be in an elevated location, with as few other land objects above them as possible.

My contention is that what happened in Colorado Springs did not go unnoticed; further tests were conducted by persons unknown, resulting in the three decades of wet weather in the United States from 1900 to 1930. The decade from 1930 to 1940 is now known as the Dust Bowl era, a time when experimenters unknown changed the experiment from utilizing copper, causing precipitation, to lead, which, being the classical non-conductor would have the opposite effect of copper,and tend to cause barometric pressure to rise and clouds to develop less readily.

This dry period didn't last an entire decade by historical accounts. I looked up the dates on the Dust Bowl, and the most intense years were 1930 to 1936. Some areas continued to suffer from drought right up to 1940 before weather patterns changed and the great plains saw more increased precipitation again. Oklahoma, Texas, all the way up into Canada experienced some drought during that decade, east of the Rocky Mountains.

The first year of the Dust Bowl was 1930, a year of drought for a semi arid region to begin with, the great plains. The poor farming practices of the three previous decades, when precipitation was abundant, left the soil exposed, and there were black sky type windstorms once drought set in. The topsoil blew away in the wind. There are now rows of trees as windbreaks throughout much of the farmland in the middle of the United States.

That wet period during the first three decades of the 20th century saw the Mississippi River overflow its banks

5 or 6 times.The Great Mississippi Flood of 1927 was the last and most significant of the flooding that occurred. Prohibition brought stills to Appalachia. Each still would have had a few good lengths of copper refrigerator tubing for condensation coils, a part of the alcohol creating process. They were usually hidden on hillsides of dense forest, meaning they wound up being located high up. The end of prohibition didn't happen until 1933, when Roosevelt signed legislation legalizing the manufacture of some alcoholic beverages. That means whiskey stills were still scattered around Appalachia for three years or more after the Dust Bowl began.

The wet weather of those three decades began before prohibition brought moonshine, but then, who knows how long bootleg whiskey was really getting made in that sparsely populated and densely forested region. It could be that bootleg whiskey reached its height at the time of the worst flooding, and condensation coils from whiskey stills were the cause of the problems of too much precipitation in the middle of the country in the first place. The middle of the United States was wet back then, but the desert southwest remained desert like. The stock market crashed in 1929 and the Dust Bowl followed the next spring.The stock market crash was due to unlimited credit expansion, and the Dust Bowl likely began because somebody decided to cash in on the stock market crash and exacerbate it with drought,selling short the whole while. A foreign country could have done something like this to strengthen the position of its own country in world wide trade. The drought began suspiciously right after the stock market crash, the very next spring.

Proving what caused this extreme change just as Americans were gearing up for even more farming after three decades of abundance brought on by increased precipitation along with new farming machinery being developed and mass produced at the time is probably impossible. I still think things happened then that everyone should have known about. The important thing to realize is that we don't need to know exactly what happened in the past, provided we have assimilated what knowledge there is to be assimilated from that time and are wise enough to use that knowledge here in the present and future.

As to how one can precisely pinpoint an exact cause and effect relationship between copper, or lead on a mountaintop or other high location and specific weather events, proof would only be inferred after numerous repeated experiments confirmed the same, or in the case of meteorological events, a similar result, in each instance. The evidence would need to be overwhelming to rule out coincidence. That just pertains to confirming that changes are caused by the inserted metal in a high location; it still leaves unanswered what, specifically, or in what combination multiple causative factors bring about the change.

Obviously the presence of the copper is causing lower barometric pressure, clouds, and rain, but how, and in conjunction with what? The electromagnetic force of the copper, much more powerful than the gravitational force, does what due to the conductivity of the copper and the effective range of the electromagnetic waves emanating from them? Could it involve the water molecules being responsive to electrical fields, static electricity, the Earth's magnetic field, the Jet Stream, particles hypothesized to explain dark matter and dark energy, or some combination thereof?

Coincidence is very possible, one couldn't infer after one experiment that a cause and effect relationship existed between copper on a hillside and a storm that passed through after three days. One would be intrigued, and repeat the experiment. Since humanity is now much more well equipped to perform such experiments and accurately document the weather events that take place than a century ago, I am confident that eventually my message will get across to enough interested persons so that eventually someone actually does carry out such experiments, and additionally, report the truth of what occurred.

The particular nature of this discovery is as much tied up with the physical properties of sub atomic particles, the province of the science of physics, or astrophysics, as it is with the science of meteorology, but that does not render any insurmountable difficulties to an aspiring student in one of those fields. A graduate student looking for an interesting subject upon which to prepare a thesis will eventually hit upon this subject and bring the matter to a proper conclusion.

There are, no doubt, obscure meteorological journals that have something on this subject in a past edition now archived. My inability to find much about the subject doesn't mean it doesn't exist. An archived article about this subject that dismisses the assertions made here, written prior to 1980, would not convince me, and it sure doesn't look good right now for the overall efficiency of science in general if one exists, for the article would be wrong.

For successful control of the weather, cooperation on a global scale would be needed, and that shouldn't be seen as an insurmountable problem either. It could be that this discovery will be the catalyst for a change of attitude of a large number of people toward other members of the species. If a problem needs to be resolved and the solution involves the cooperation of the entire population,such an undertaking might help to form more comfortable relations between different people of the entire world.

There is friendly competition between nations, and that could extend to some kind of gardening competition. A contest that would depend on optimal weather for the victor to claim the prize could begin, and each participating country trying to win the prize acting together accomplishes the world wide goal of distributing water plentifully and safely to all land areas. Correspondingly less would be left to the oceans to generate hurricane/typhoon/cyclone activity.

Island chains eventually look after themselves, in so far as arranging for precipitation to occur. On some of the resort islands where the lifestyle is much more relaxed, some of the residents may begin experimenting with making clouds. Different quantities of copper in varying locations on islands produce different clouds, lead would cause the clouds to disperse differently depending on where it is, and when thunder and lightning is occurring, some fireworks could be added to the display, making island sunsets that much more spectacular. Some of the cloud makers could become adept at making clouds appear that resemble objects roughly, at least for a short while, and as the years progress, island life changes to accommodate the new sport of cloud making as an art. Sunbathing tourists get a free cloud art display, with clouds disappearing completely at times followed by unusual shaped clouds, a thunderstorm with fireworks added at times, clouds suddenly appearing on one side of an island and then vanishing as they cross to the other side, etc. The cloud view might be an added feature on resort islands in the future.

Water is precious, but it is also abundant in large quantities. The only problem facing mankind is the relative distribution of the water over the land masses of the planet. Placing copper tubing in a propitious location at strategic times could so cheaply redistribute fresh water around the planet and prevent extreme weather from destroying things, provided the entire world is intent upon an ample supply of water,that it is imperative that not much further time be wasted taking this idea forward. Desalinization plants are definitely not the way to go.

The atmosphere has been assumed to be comprised of electrically neutral particles, not affected by the electromagnetic force of copper, leaving only the gravitational force to influence the tiny components of the atmosphere,a force that is 100 billion times weaker than the electromagnetic force. The stable isotopes of nitrogen and oxygen, N_2, and O_2, are electrically neutral and that is 99% of the known air, and most of the remaining one percent is also neutral. A path of least resistance created by a quantity of copper on a mountaintop wouldn't seem at first glance to be likely to have much of an effect on the atmosphere, and consequently, the weather, at all.

However, the water molecule itself is not electrically neutral. It is a polar molecule. A book titled "Giant Molecules" by Alexander Yu. Grosberg and Alexei R. Khokhlov on page 43, 44 and 45 discusses the nature of water. Some of what was mentioned follows; "The water molecule is triangular in shape. The electron cloud tends to be shifted away from the hydrogen nuclei toward the oxygen nucleus by (a certain amount in an equation). As a result, the positive charge of the hydrogen nuclei is not quite compensated. Similarly, there is an uncompensated negative charge around the oxygen nucleus. This peculiarity of the structure may not seem of great significance at first sight. However, it is the real cause of all the special properties of water that make it play such an important role in living organisms. A water molecule has a considerable dipole moment, so the molecule is polar. This means that in an external electrical field water molecules can be regarded as little "dipoles," each carrying two charges, +E and -E, separated by a distance. Such little dipoles have no difficulty in becoming aligned in an external electrical field; this explains why the dielectric permeability of water is much higher than for all other common liquids."[3]

The statement about alignment in an external electrical field is referring, no doubt, to water in liquid form in some quantity, in a container. All the liquid water molecules in the container would have the hydrogen pointing one way, and the oxygen the other, when exposed to an external electrical field. I see no difficulty in jumping from those observations to the observation that a water molecule in gaseous form would react to the type of external electrical field created by the presence of the Tesla Coils on the roof of that gentleman's laboratory in Colorado Springs during the second half of 1899, as it meandered freely around the planet amongst other gases. Since water molecules are very weak magnets, using a huge magnet similar to what junk yards use to lift junk cars by placing it somewhere high up and activating the electrical current might do an even more efficient job of herding water molecules to a desired location. A utility company could have a little fenced off area near a mountain top with a small gas generator connected to a huge magnet there. When precipitation is required, someone activates the generator, providing the magnet with electrical current. After desired precipitation levels are reached, the generator is turned off, and the magnet is no longer pulling water molecules toward it.

Reverse the direction of the current, and the negative charge will repel the negatively charged oxygen nucleus of each water molecule. Since the oxygen nucleus in a water molecule is larger by far than the two hydrogen nuclei, those two hydrogen nuclei are always along for the ride, the larger component deciding where the molecule travels. Thus, positive or negative charges from a huge electromagnet in a high location could mimic somewhat purified copper or lead deposits similarly located. Proof would still be needed to satisfy scientific rigor. Now, at least, there is more than one mechanism to be experimented on. A path of least resistance might not be necessary with the use of huge magnets instead of copper in the hundreds of pounds. That remains to be seen.

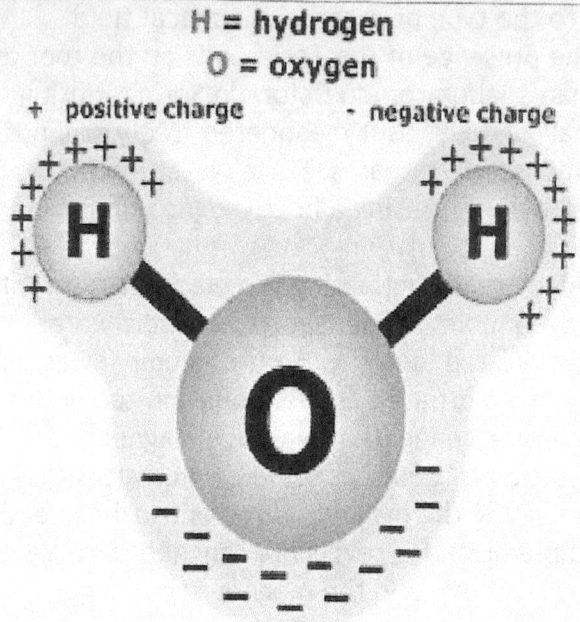

H = hydrogen

O = oxygen

+ positive charge   - negative charge

The larger Oxygen nucleus decides the path of the water molecule

The atmosphere varies in how much water vapor is present at any one location, and by itself, may not be enough to account for all the effects observed when copper is placed in a high location and left there for some time. The slow increase of water molecules over 72 hours would raise the relative humidity, and with more and more water molecules present, all traveling in the same direction, the electrically inert nitrogen and oxygen pairs get pulled along with the water molecules, and barometric pressure falls as a result.

Rather than to just pin all the blame on the water molecule, the next chapter deals with what may be accompanying the water molecules along the path of least resistance. After all, the scientific community appears not to consider the properties of the water molecule itself to be significant enough to cause any changes in the weather, since this process I maintain exists isn't in encyclopedia.

There really isn't that much water by weight in the air when the saturation point is reached. 100% relative humidity at 30C is 3% of the atmosphere by weight. Naturally any precipitation counts as something more than 100% relative humidity. The usual 99% of the known air, nitrogen and oxygen paired isotopes, can fall to 96% of the known air if relative humidity is at its height at this temperature. For barometric pressure to fall, and all air bound entities to travel in the same direction, more than just water molecules are herding along the path of least resistance. These additional entities add more weight to the water molecules following the path of least resistance, those water molecules in and of themselves not comprising sufficient mass to bring the electrically inert paired isotopes of nitrogen and oxygen along gravitationally.

Chapter 2. A Cosmological Explanation

" Everything Is Not Now Unbroken"
From A Song

There is the likelihood that whatever fragments the Big Bang created have been condensing into hydrogen molecules all these billions of years and may still exist, comprising the missing energy and mass that different scientists have been searching for. The smallest thing in the universe, the most abundant thing in the universe also, combines to create hydrogen atoms, theoretically, through a process we will discuss later.Dark matter and dark energy are hypothesized because the universe is expanding at an increasing rate and with the total mass of stellar objects known, there is not enough mass to slow the expansion and eventually bring it to contract back upon itself gravitationally, creating another big bang.

There must be more matter and energy prevalent that is undetected, then, according to current theorists. Other evidence suggests it is not just theory. The orbits of the celestial bodies are taking place at much higher speeds than they would if what is observable were all that existed. The calculations of the astronomers and astrophysicists indicate that 74% of the universe is dark energy, 22% is dark matter, and the remaining 4% is the matter and energy already discovered in the known universe. This may seem a little confusing at first since matter and energy are inextricably intertwined, but astronomical calculations do indicate that more mass and energy must be present. The terms dark matter and dark energy have been employed by the scientific community since whatever it is that they turn out to be, they have not been detected as individuals as yet, and have thus been termed dark. One of the first published articles about dark matter and dark energy was from as far back as 1916, but little attention was paid to the idea until the last few decades.

Energy separates from matter when it is released, as in a nuclear fusion event. Energy is either released from matter, or is potential energy that could be released from matter. We would assume that the missing energy and mass are in the form of things that are a combination of both, not all energy being released from any current matter. When nuclear fusion occurs between two hydrogen atoms, a lot of energy is released and the two hydrogen atoms fuse into one helium atom.This helium atom still has potential energy within it. It can also undergo nuclear fusion with another helium atom, release a lot of energy and become a still heavier element.

The two things known as matter and energy are arguably two sides of the same coin. Actually distinguishing when something is energy and when it is matter is not exactly easy. Matter is used to describe things when one measures the gravitational effects of something. A star could explode in a supernova and what was matter and energy is now predominantly energy, so the coin has now flipped sides, although gravitational waves are still propagating from every atom. Now it is energy,mostly. Eventually, a new solar system with planets, moons, asteroids, comets and a smaller star at its core will coalesce, and the coin will have flipped back to some less energetic combination of matter and energy.

Energy refers to what is released from matter. In nuclear fusion, in supernovae, energy is released from matter. Even a pile of leaves on fire releases energy, heat. On the cosmological scale, energy also relates to the motions of things. The faster something is going, the more energy it has. The greater the energy, the greater the effect of seeming to pull the known universe apart. The elements, the matter that is known to us, can vary in how energetic they are. Five pounds of something at rest has much less energy than five pounds of something moving near the speed of light. The only problem with that is we don't really know where "at rest" is. It is a term usually meant to refer to those things in one's vicinity that appear to be motionless, yet are moving at quite a high speed right along with the observer. More on space and our position in it later.

If it turns out to be one elementary particle from whence hydrogen originates it would then replace hydrogen as the most abundant thing in the universe, by far. A book by Dan Hopper, Dark Cosmos: In Search Of Our Universe's Missing Mass And Energy[4], has some interesting parts. When discussing possible candidates for the missing particle the author says it couldn't be a charged particle, because particles with charge would interact with photons, making them luminous, and thus detectable. But can he be absolutely sure that things in the world of particles too small to detect are taking place as he thinks they are?

An electromagnetic wave coursing through the air from a quantity of nearly pure copper may do something entirely unexpected to particles that small, and they might still remain non-luminous. The search for truth would have to leave room for unexpected possibilities.A particle that is not charged, but behaves like one in the quantum mechanical universe, or a charged particle so minute that the interactions one would expect it to undergo when encountering a photon go undetected, for example, couldn't be ruled out.

Magnetic monopoles have been hypothesized to perhaps have been created by the big bang,as numerous as protons, though it is assumed they would have huge amounts of energy. Some Grand Unified Theories have hypothesized that a transition took place early in the history of the Universe, in which the three forces of the Standard Model, the electromagnetic, strong and weak forces emerged from a single grand unified force. "As a consequence of this process, an enormous number of strange objects called magnetic monopoles would be generated."[5]

Magnetic monopoles appear to be the most likely candidate. These are magnetic particles with only one side of a traditional magnet. They are not found in nature in the sense that any magnets discovered or created have always been found to have two poles. This refers to objects large enough for humans to see and touch, and any magnet such as this when cleaved in two, will be two magnets, each a dipole. If the primordial specks were magnetic monopoles each one would be energetic and unstable, definitely a charged particle.

According to the prevailing theory of the origin of the Universe the Big Bang occurred some 13.6 billion years ago, and all matter began flying apart at an incredible rate. Eventually, the hot plasma, the scientists say, cooled sufficiently for hydrogen atoms to begin to condense. The hydrogen collected in large quantities gravitationally to form stars, the stars began to burn hydrogen via thermonuclear fusion, creating helium, the next heavier element. Eventually some stars proceed to burn helium in thermonuclear fusion, and some may even have fusion going on between yet heavier elements.

When a star burns up most of the fuel available to it, gravity causes it to collapse upon itself, then explode in what is called a supernova. These supernovae are the source of elements heavier than hydrogen and helium due to the tremendous pressure and heat generated in these events. Hydrogen moves on to become around 96 or more different naturally occurring elements, through these two processes. It has never been explained what condenses into hydrogen, or if that process is finished now. All we get is that hydrogen condenses out of the plasma created after it has cooled sufficiently.

There have been books detailing how protons condense out of the plasma at such a temperature and density, neutrons condense out at a different temperature and density, electrons at still another temperature and density. One of the books I read recommended a book by the physicist Weinstein. The crucial question,though, is do these now cooled entities naturally condense into hydrogen all by themselves? If that were so, one would expect quite a bit less dark matter and dark energy to be prevalent currently. The huge quantity of dark material must be the breeding ground for more hydrogen, in which case, the idea that the cooled plasma condensed spontaneously into hydrogen long ago and is finished now seems wrong.

I just doubt that exploded fragments from a huge explosion would be something other than precursors to hydrogen. The primordial soup mentioned on science programs refers to the Oceans of the Earth in the early years of its existence. The known universe could be viewed as so much primordial soup at an early time in its existence as well. Hydrogen appears to be the only thing making any progress in the primordial soup of the early universe, and a lot of the universe appears to still be whatever fragments originally created in the Big Bang, so it seems more likely that hydrogen in the known universe is still on the increase.

Plasma won't remain plasma after it has cooled. Hydrogen is the starting point for all the other elements. The early universe was blown to smithereens, then, and the plasma the scientists refer to once cooled would be precursors to hydrogen, of an unknown constituency, primordial specks. The most pulverized thing possible, the smallest remaining unity of energy and matter that remains after the colossal explosion of the Big Bang strips everything away and into space, a magnetic monopole.

The entire process of hydrogen creation may be a very long and still ongoing process. One hydrogen atom by itself is the tiniest atom in existence, and for these to come into existence one at a time, without, of course, announcing themselves, how would we know? We

wouldn't be aware of the process if it were ongoing.
Why every book I've ever seen that touches on this
subject makes the assumption hydrogen production in
the known universe ended long ago is beyond me.

Figures for the relative distribution of hydrogen in the
known universe give us 93% of the known elements, by
far the most abundant. Yet that 93% is only 93% of the
known matter in the known universe; there is also the
96% of the known universe that is dark matter and dark
energy. Distribution of hydrogen on rocky planets like
the Earth tails off considerably, the Oceans have 11%
hydrogen, other areas less, 3% in the Earth's crust.

Hydrogen naturally pairs up as stable H_2 isotopes.
These two hydrogen atoms could prove to be a
reproducing pair, in the sense that a spontaneous
metamorphosis may take place between the two. The
double combination of stable hydrogen isotopes serves
as the template, it provides, in the space between the
two nuclei, a residual image into which scores of
primordial specks swarm, somewhat like stem cells.
Wherever they happen to wind up in the template, they
adopt the role of that part of the hydrogen atom. Some
could wind up as part of the electron, others a part of
one of the quarks in the nucleus. The electromagnetic
and gravitational waves leaving the two identical
hydrogen nuclei are very strong forces at such small
distances, strong enough to imprint new information
onto the primordial specks.

Enlarge the entire scenario of a pair of hydrogen
atoms until the nuclei are the size of a grain of salt.
The distance separating the two nuclei would have
grown to three hundred feet or more, and the two

electrons would be 150 feet or more away from each nucleus.From that perspective it is evident that there is ample room for tiny particulates.The nucleus of a single hydrogen atom is 1.75 femtometers. The electron cloud surrounding a single hydrogen atom is 145,000 times larger than the nucleus itself. That makes the electron cloud slightly larger than one fourth of a nanometer.

Enlarge the nucleus one million times, and it is 1.75 nanometers, still invisible without a microscope. How big the electron orbital would be once the nucleus reaches the size of a grain of salt would depend on the size of a grain of salt. If the electron orbits in a path 145,000 times bigger than the nucleus,that would mean that the orbital path of the electron is 145,000 grains of salt in diameter.

If there are 24 grains of salt to an inch, the electron would be in orbit in a circle slightly more than 500 feet in diameter. At 40 grains of salt to an inch, the electron orbital path would be slightly more than 300 feet in diameter. A grain of salt in a bubble nothing can penetrate at least 300 feet in diameter. All that space for one grain of salt, with an electron 2840 times smaller circling. Two hydrogen atoms together in a paired isotope have separate nuclei, each with one electron circling, though the two electrons could switch nuclei and do figure eights simultaneously, I think. Each nucleus would have a protective electron shield preventing the other from coming any closer. The two nuclei would be a minimum of 300 feet apart, and a maximum of more than 500 feet apart, with the nuclei the size of a grain of salt.

Again expand the whole scenario until you are

standing between the two grains of salt, and there is a pounding rhythym in the air, like huge printing presses biting down on reams of paper, as each pulse emanates from the two nuclei. Each nuclei, with its tightly encased trio of quarks locked in gravitational battle and spinning rapidly, emits pulses of electromagnetic and gravitational waves. Standing in the middle of the enlarged hypothetical scenario these pulses could even be deafening, if one could actually enlarge the situation accurately, and the resultant waves strong enough to seem like the shock waves from a nuclear explosion. None of that is really possible, though, but it does give us an insight into the possibility of hydrogen reproduction by the stable paired isotope of hydrogen.

Space being so much more prevalent than the tiny particulates themselves makes it evident that many primordial specks could actually accumulate in the space between the two hydrogen nuclei of the stable isotope and therein acquire new characteristics from the emanations from both sides occurring there. It appears to be the only location in the known universe where information could be transferred from existing hydrogen atoms to the primordial specks created in the Big Bang.

Taken at its actual size, the electron shield provided by the rapidly circling electrons around a pair of hydrogen atoms may very well prove insufficient at keeping everything from penetrating the area around the two nuclei, since hypothetically dark matter and dark energy components could be small enough to enter that space. Two hydrogen atoms together at actual size, each surrounded by an electron cloud slightly more than a quarter of a nanometer in diameter, would be about that far apart from each other, the distance of the radius of each orbiting electron. Something in the space between the two nuclei would be around an eighth of a nanometer from identical atomic nuclei on each side.

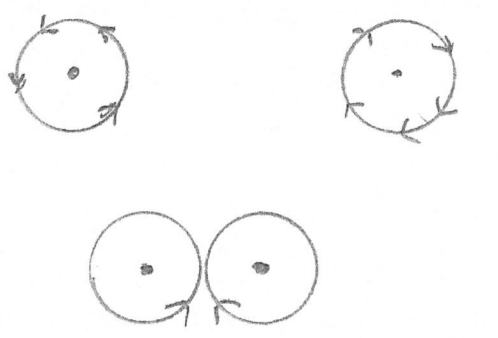

Two H pair up. Where the electrons meet, new hydrogen is made

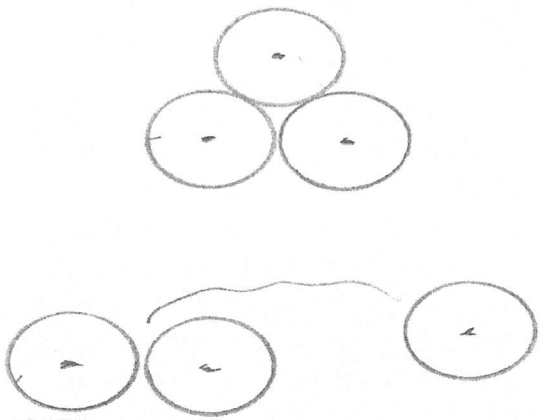

Deuterium is created, and soon separates, being unstable

For the sake of argument lets suppose magnetic

monopoles about one tenth of a femtometer in diameter, each hydrogen atom requiring around twenty of these things to coalesce into a single hydrogen atom. These tiny things all by themselves, with no electron, about one eighteenth the diameter of a hydrogen nucleus, could conceivably have accumulated in that space between the two hydrogen nuclei by the thousands when the universe was much younger.

Currently, in stars, 93% of the known matter of the star is hydrogen, and gravitational forces being much greater in stars than on Earth, primordial specks would be in greater abundance. Hundreds of primordial specks could be, at any moment, jockeying for position to be those that wind up in the template to become the next hydrogen atom, in a huge number of paired hydrogen isotopes.

Hydrogen is a stable atom, the simplest atom. The most pulverized thing possible existing after the Big Bang isn't going to know how to combine into one; the plasma of the known universe, once cooled, quite soon came into contact with some H_2 isotopes from somewhere else in the much larger entire universe, and that encounter set the process of hydrogen production in motion by providing a template for the magnetic monopoles to transform into hydrogen. Ever since that point hydrogen production would have increased as more and more templates emerge over the eons, as single, newly created hydrogen atoms pair up.

One would naturally wonder where the H_2 came from in the much larger universe to start things along. Somewhere,at some time, hydrogen must have occurred spontaneously.It could happen to emerge spontaneously to some extent in the known universe shortly after the plasma cooled. However, since hydrogen must still be materializing without our perceiving it, spontaneous hydrogen production may have only been possible when the plasma once cooled was still very dense. Further hydrogen nucleo-synthesis currently requires paired hydrogen isotopes as templates.

If all that were true, a strong electrical field running through a large sealed chamber with a few hundred hydrogen atoms within to serve as templates could prove that hydrogen is still coming into existence if left sealed for several months or even a year. Opening it after some time should see more hydrogen atoms within than previously. The primordial specks would accumulate along the electrical field, pass right through the walls of the container, encounter hydrogen pairs to reproduce within, and produce more hydrogen. One would, of course, have to try a completely empty chamber under the same conditions as well. What is eventually found there could also prove instructive.

The presence of stars and other celestial bodies changes the environment somewhat for the hydrogen atom; there are now places where there is little hydrogen, and places where there is a lot. The entire body of stars would have lots of hydrogen templates. The atmosphere of Earth would have as many H_2 isotopes as the relative humidity in an area. Each water molecule is one half hydrogen factory, under the right conditions. The centers of rocky planets and moons probably wouldn't have many templates for hydrogen production, especially cold ones like the Moon or Pluto. The lighter elements would have made for the surface long ago, or be molecularly bound to something else.

All the water on Earth, each water molecule containing the hydrogen template, could be a vast source of new hydrogen. New hydrogen atoms hypothetically created in the sea would rise to the atmosphere also, while pairing up, most escaping into outer space, some finding an oxygen atom to join with on the way up and out. The oceans would be making more water vapor than mere evaporation would account for, so more would accumulate in any one place if experiments with copper in a high location were attempted. There is simply more moisture available than previous estimates.

To better estimate atmospheric events, the totality of water vapor being additionally created by hydrogen templates creating hydrogen that eventually become a part of water molecules themselves would be worthwhile knowing. Perhaps a computer simulation with an additional two tons of water daily programmed to begin to exist could give a more accurate depiction of available water vapor. We may already know something about how much hydrogen is getting created in the Oceans. After all, if that type of activity were going on at a very rapid rate, the Earth would already be almost completely covered in water, so the rate that hydrogen is coming into existence within the oceans probably isn't very rapid, unless the rate at which H_2 escapes into outer space is very high. That would mean that concentrations of primordial specks here on Earth probably isn't high enough for hydrogen to reproduce rapidly.

The Earth gains tons of mass daily, as small meteors

are vaporized to dust upon entry into the atmosphere. At least that is where this weight gain comes from, according to scientists. This daily gain of mass could be as much as 100 tons. Would anyone be the wiser if one or two tons of that total were newly created water molecules from newly condensed hydrogen? We would not know whether some of the daily weight gain came from there or not. Measurements of the Earth are taken from space, no doubt. If both dust accreted on the surface and hydrogen was created within the sea,paired up and became water molecules upon contact with an oxygen atom, amounting to two tons of water daily, the Earth would just appear slightly larger, and that it does.

Each day adding a ton of new water to the Earth through reproduction of hydrogen, water being where some of it would end up as,would compare to putting a teaspoon of water in a lake with a surface area of a square mile on a daily basis. A gallon of water is around eight pounds, thus a ton of water would amount to around 250 gallons of water. Even two tons of water, 500 gallons,wouldn't be noticeable at all spread out all over the oceans on a daily basis. Hydrogen has to combine with oxygen to become a water molecule, so 500 gallons of water isn't all new hydrogen adding to the total weight of the Earth.The oxygen in the atmosphere that combined with hydrogen would have been present in the atmosphere, but it now adds to the mass of the Earth as water on the surface, eventually.

The nuclear furnaces of stars would be the place where most hydrogen reproduction would be transpiring. Unfortunately, that makes examining such an event closely impossible. Stars being larger than Earth by far, gravitationally more primordial specks would be bound to a star;the density of those being the critical factor in the presence of hydrogen templates. In places with weaker magnetic fields the density of the primordial specks would ordinarily be below the required threshold for hydrogen to reproduce, or as hypothesized in our case, diffuse enough to slow the process down. The gas giants in our solar system are probably above the threshold even more so than Earth, so one would expect more of that to be happening on the gas giants, continually adding to the gases present, with condensation eventually adding to the land mass. Perhaps a space journey to one of the gas giants a hundred years from now will be when and where hydrogen reproduction is conclusively proven to be still ongoing in the known universe.

Primordial specks surround or are within whatever objects of mass are in the universe. Some of these primordial specks may be in deep space,but most would be gravitationally bound to something massive. All of the celestial bodies would also have some of this astonishing amount of undetected mass and energy within them, since the extreme tininess of these things would allow them to pass right through solid objects. With regard to black holes, the primordial specks would not be so much around them as within them, and once that happens they are no longer what they were.

On the surface of the Earth, primordial specks would be in great abundance, a veritable blizzard of them.The extent they reach into outer space around the Earth could be as far as hundreds of miles. If this staggering quantity of dark matter and dark energy are magnetic monopoles they would indeed add to the water molecules following the path of least resistance.

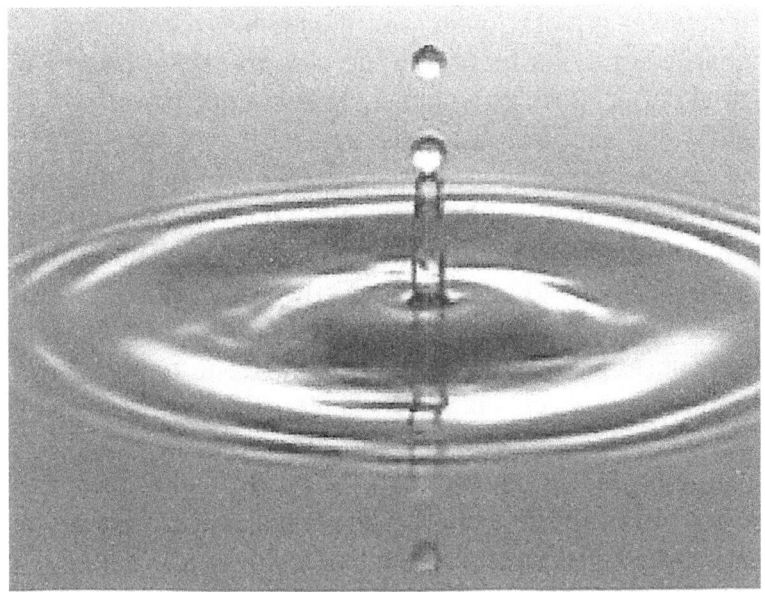

A drop in the bucket!

Since the initial fragments of our proposed hydrogen precursors each contain mass, albeit in an exceedingly small amount, the combined mass of all the fragments accumulating along the path of least resistance created by a quantity of copper strategically placed could exceed considerably the mass of the air itself. At the highest estimate of 96% of all matter being dark energy and dark mass, and assuming the Earth has around it a proportionate share of primordial specks, and increased concentrations of primordial specks along the path of least resistance increasing that quantity by a factor of two, for example, the dark matter and dark energy going with the atmosphere could be as much as forty eight times as much matter and energy as the atmosphere itself.

Therefore, these small entities would influence the air molecules gravitationally, air molecules that are not charged,and unaffected by the electromagnetic force of the copper. Besides which, any object not traveling along the path of least resistance with everything else would be continually bombarded by these tiny objects, though this may not be much of a factor because of the minuteness of the primordial specks. The smaller of the two things would be deflected. The water molecule is much larger than the hypothesized primordial specks,so encounters with it could occasionally tend to position other atmospheric components along the path of least resistance.

The newly formed hydrogen atoms that may have begun to coalesce from the increased concentrations of primordial specks and water molecules would quickly pair up as stable H_2 isotopes,and rise, being the lightest element, even if now paired up. As these newly condensed paired hydrogen atoms rise, the moment one of these encounters a free oxygen molecule it joins with the oxygen molecule and becomes H_2O. Ozone,O_3, is continually being created and destroyed as oxygen atoms are lost or gained in the upper atmosphere by stable oxygen isotopes, O_2.

Clouds appearing where none were expected would be explained by this theory. The primordial specks probably reach concentrations sufficient to start synthesizing hydrogen atoms inside the H_2 in the water molecules more rapidly than the usual slow rate.These new water molecules join clouds from above, while currently existing water molecules in the atmosphere from hundreds of miles in every direction begin accumulating along the path of least resistance below, and rise to the clouds. H_2O is lighter than the stable pairs of oxygen and nitrogen.

There are other possible contributing causes for the effects perceived with copper creating a path of least resistance. The Jet Stream could change course as a result of a row of copper tubing changing the magnetic field of the Earth to some extent. It could be that the static electricity following the path of least resistance weighs more heavily than expected. If the copper attracts static electricity to it, there is correspondingly less static electricity in the air surrounding the copper, and that could possibly make the path of least resistance easier to follow for electrically neutral particles. Perhaps the correct answer to the question is "all of the above". Each little thing by itself is maybe not sufficient to cause weather changes by itself, but is a part of the equation. The combined effect of all the mentioned possible causes produces weather changes.

In my opinion, a large percentage of the effects observed are due to the polar nature of the water molecule, and all the primordial specks and the astounding amount of mass they bring with them along the path of least resistance, in addition to the hydrogen that comes into being from the increased concentrations of these.If one were to pie diagram the possible causes, I would give the polar nature of the water molecule at least 40% of the pie, the primordial specks 30%, the Jet Stream 15%, static electricity 10% and 5% to the Earth's magnetic field.

Best we will ever be able to do in regard to the possible causes is guess. It is clear though, that despite some doubts as to the exact causes of the increased clouds and precipitation, nevertheless a half ton of copper on an 8000 foot mountainside near an ocean left in place for 72 hours would provide evidence all by itself, especially if one chose to begin the experiment when no clouds or precipitation were expected in the next few days by meteorological forecasts.

Chapter 3: More Cosmology

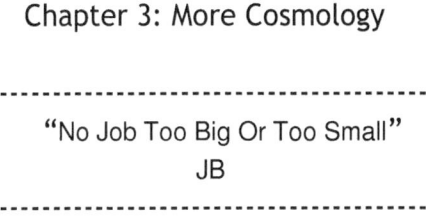

The known universe is speeding up in it's expansion. Common sense tells me that is because there is more matter beyond the known universe exerting a gravitational pull on all that is the known universe. This speeding up has gotten a lot of speculation in the astrophysical, astronomical, and cosmological scientific communities.

I don't know how exactly this was determined, but astronomers maintain that at one time the universe had been slowing down as galaxies careened farther away from each other, but that turned around into accelerating expansion some billions of years ago. Almost all cosmology books written give possible theories as to why this is happening, none but this one, at least that I have read, supposing simple gravity to be the culprit. The idea that the known universe is not the entire universe, and that matter and energy at a distance of 15 billion light years might not be detected since the light is too faint suggests itself. A guess would give us a possible ratio of how much dark energy and dark mass are local,and how much is matter and energy more distant if one took the totals we have currently.

74% is dark energy, which is the force trying to pull the universe as we know it apart, 22% is dark matter, holding our known universe together, and 4% is currently detected matter and energy, and since matter and energy are both contained within matter initially, and we have 22% dark matter holding us together we probably have a proportionate share of somewhere around 22% dark energy trying to tear us apart in the local arena, and the other 52% dark energy would consist of the pull of distant objects upon our ever expanding Big Bang. These distant things would be objects with matter and energy combined, just like in the known universe. That still leaves us with an 11 to 1 ratio between local dark matter and dark energy, and the known elements, neutrinos, etc.

"Mommy, why doesn't common sense say anything to me?

Astronomer Vera Rubin spent a great deal of time measuring the speeds of orbiting stars. She published a paper describing her detailed observations of the motions of stars in the Andromeda Galaxy and her conclusion was that "for the stars to be moving with the velocities they had, there would have to as much as ten times more mass in the galaxy than was visible."[6] The difference between 10 to 1 and 11 to 1 is only around 1.5%, surprisingly small. Her exact words were ten times more mass, not mass and energy, but one could take a wider view and assume she meant things, a combination of both matter and energy, since that is what exists. Or one could assume a corresponding amount of dark energy to equal the amount of mass.

Other than those figures, there is the consideration that our known universe is still in a rather energetic state; possibly more dark energy is present here, and less attributable to distant massive objects. Lots of energy is being released in the stellar activity of 100 billion galaxies. Add to that the energy released in supernovae, and a good argument could be put forth that the gravitational pull of distant objects accounts for less than 52% of the total dark energy that we have yet to find.

Stellar activity and supernovae are not dark energy at all, they are quite known to us, but one could impute more energy to dark matter and dark energy, if one were to suppose them to be hydrogen precursors. Once hydrogen, they are then capable of participating in nuclear fusion. Nevertheless, something is causing galaxies to fly away from each other at increasing speeds, and to clump together in enormous clusters and leave huge empty places in the known universe. The observations of Vera Rubin point to around a 10 to 1 ratio of primordial specks to known matter and energy, and the proportionality guess gives us an 11 to 1 ratio, so perhaps the figure for dark energy here in the known universe could in reality be 2 or 3 percent higher.

The last few years have seen two surprising discoveries that are still being investigated. In 2006, NASA scientists launched a balloon with some sensitive detecting equipment to more closely examine faint radio signals from the most distant stars. What happened was an extremely powerful radio signal from somewhere far off totally drowned out any hope for them of recording from the faint stars they were trying to understand better. This radio source was 6 times more powerful than all the radio galaxies in the entire known universe. What this radio source is can't be on too long a list. No doubt it will eventually be concluded to be a big bang in its own right, the light having come and gone, or too faint to see, but the radio waves very strong. I found this article on the internet with the title "NASA Mystery Boom; Or Something in Space Is Screaming". I hope that soon, more will be available about this singular discovery. I'm assuming it has not stopped broadcasting powerful radio waves.

An astronomical team led by Andrew Kashlinsky published an article in 2008, detailing what came to be referred to as "Deep Drift". There is a point in space, according to these astronomers, where the things in the known universe are being pulled toward, those galaxies closest that point in space traveling toward it at extremely high speeds, as much as three million miles per hour, those furthest away much slower.It seems the entire known universe may be already in the grips of a gargantuan black hole. I found this article on the internet as well. If the reader were to search for "Andrew Kashlinsky Article", I'm sure they would find the one about "Deep Drift". Neither of these two articles were very long or greatly detailed, at least the news articles. Something being discovered possibly from outside the boundary of the known universe, two different such possibilities, has implications in Cosmological theory.

Both of these recent discoveries were made prior to the release of the first edition of this book. Without a doubt, the author can be shown to be human,and admit of limitations. I didn't know that a plurality of discoveries had occurred involving possible entities from beyond the boundaries of the known universe, and was similarly unaware of the polar nature of the water molecule when the first edition was released. All of these things I didn't know all support my theory and strengthen it. That I was capable of perceiving that changes occurred involving the atmosphere and copper in the hundreds of pounds in a high location without knowing of the polar nature of the water molecule, and also that the first edition maintained throughout that the entire universe is larger than the known universe, just goes to my powers of observation.

Almost all the theories in Cosmology start with the assumption that the known universe is the entire universe, that space/time is a fabric of some kind, and that there is something called vacuum energy out of which particles pop out.Could it be that a much simpler explanation might be the truth?

Massive objects could still bend light waves even if space/time is nothing other than what they are.Photons have been found to have no mass but they are definitely something, which came from something with mass, a kind of energy, so why do we need a fabric? They are pulled by gravity. I cannot elaborate what the force is that is called gravity. Newton evaded the same question. How a force can reach through empty space is beyond me, yet common sense tells me that is what is happening. Space doesn't change, a force moves through space. It makes the most sense that it would be a wave traveling at the speed of light, which is the currently held view.

Particles don't pop out of vacuum energy, they synthesize from smaller ones. The recent findings that the vacuum contains energy by some scientists seems to overlook dark matter and dark energy completely, which may be responsible for the figures these scientists are arriving at. To conclude that an area of space is a perfect vacuum when the unresolved question of what and where dark matter and dark energy are seems to be looking blindly at the vacuum.

The more distant an object is from us, the faster it is receding from us because the more distant it is, the closer it is to undetected massive objects, hence the more distant objects in our known universe are more greatly influenced by the gravity of these other massive objects hitherto undetected. For the entire universe to have been contained within the Big Bang and now, after almost 14 billion years, the entire collection of 100 billion galaxies is not slowing down as it careens farther and farther away from where everything in the known universe was at the Big Bang makes no sense at all in the framework of the natural laws as we know them.

Gravity would have begun to slow the expansion eventually, so now theorists are inventing repulsive gravity to explain the excess of dark energy. How about normal gravity, just elsewhere? We have neighbors on the fringes of the known universe. The mass they contain could round things out. Therefore, the Big Bang most likely didn't happen as theories now state. In fact, exactly what we are observing would be the case; dark energy would be much higher than dark matter if a considerable amount of matter were too far away to detect, yet were exerting a gravitational pull.

If we throw away those three assumptions, and in the words of Henry David Thoreau, simplify, simplify, simplify,we perhaps answer these questions, at least as satisfactorily as we can given our limitations.From what I can gather there is a problem in physics concerning uniting the gravitational force with the Grand Unification Theory, or with relativity, or quantum mechanics, and perhaps these new insights into the probable location of the excess 52% dark energy, and also where the missing particle can be found to concentrate which makes up probably somewhere between 90 and 92% of all the matter and energy in our known universe, and is the primordial speck, can help the mathematical geniuses in the world solve that puzzle.

At present, more and more exotic theories are emerging to try to bring successful answers to questions such as what happened in the first few seconds of the Big Bang, why the known universe is now speeding up in its expansion, and where dark matter and dark energy are. The complexity and unusualness of these theories is an indication that some simplification is overdue.

The notion of space/time as some kind of coexistent

fabric came about originally as part of Einstein's Theory of Relativity. That theory and its predictions would all probably still hold true even if space and time turn out to be different. From the beginnings of language,events and things were perceived in the real world to take place and have what one would call duration. We now have many words to indicate various amounts of time. Something obviously changes from when an egg is raw to when it is cooked, beside fire being applied to it.

The fire is applied only so long or the egg is burnt. Actually describing the passage of time is difficult without referring to it somehow,as in the "only so long" referring to how much time passes. These same events, to be perceived in the first place, had dimension also,in effect they were large enough to be visible to human eyesight. Beside that, there was certainly a lot of space around. Even if the ancient philosophers didn't have quite the astronomical knowledge we do now, space was everywhere, distances were calculated, maps were drawn to scale.

Philosophers mused over space and time quite a bit way back when, and decided time and space were both immutable absolutes, eternal, unchanging, or something like that, others claimed they didn't really exist. The problem with both space and time are that they are not tangible things, comprised of known elements. Space not having any features other than being a completely frictionless medium through which any and all matter and energy can travel, is it something? There is an up, down, left and right direction anywhere we have found. I don't think it really is anything, the dimensions of space are mere emptiness. One would have to build a wall to impede movement.

Time has a unique kind of existence, as well. The universe has things in it, they are traveling through space, and inescapably, one moment continually gives on to the next. It would seem there would be no other way for the things in the universe to travel, but for movement to yield a new destination after the lapse of time. Time can't not exist, but it isn't really anything but what must be there to permit the movement of things, just as the dimensions of empty space. We see that as a thing moves, it goes from one place to another. Space and Time couldn't not be there for that to happen, therefore they could both simply exist as a consequence of there being moving things. Rather the empty space is there and the lapse of time just a necessary consequence of things in motion.

A space ship leaving Earth, traveling near the speed of light for 100 years upon then returning to Earth will find family and friends long dead, though only about five years would have elapsed for the inhabitants of the spacecraft. This tells us there is theoretically a variable factor in time, though not a very practical one for human intents and purposes. Time nonetheless appears to be ever moving forward, slower in the space craft than on Earth.

Space is where all this takes place, the spaceship would have traveled an incredible distance, but could space alone still be an immutable absolute,in short the edges of the universe do not exist, for there is no way for space to not exist? Why did time contract in the spacecraft, were space not really tied to time? Travel speed seems to slow down time, and it takes extreme speeds to do that to any extent. It happened in space, true, but speed of travel, not variability of space, is what caused it. Even the idea that velocities near the speed of light slow down time could be based on faulty assumptions, since it is as yet only theoretically possible, and almost impossible to prove. Movement, not space, may slow time down some.

It would be difficult to imagine that there might be a far location inaccessible to navigation, if one had some device for instant teleportation. Space thought of this way could give us a clearer picture of things,since if the void stretches forever calculations change. If time alone ultimately admits of changes, together with some figures to calculate total hidden mass and energy, that may make some computer simulations come up with a more accurate picture of the universe.

There was a solar eclipse in 1919 where Einstein predicted that the light from stars that are behind the Sun from our vantage point would bend as they passed by near the Sun, and astronomers confirmed that this is what actually happened. This observation was one of the things that served as the basis for the space/time continuum idea.

Yet the crucial question we must answer is does space really need to undergo some kind of change for this event to take place? Could it be that space, eternal, infinite in every direction, stays the same and the photons change course? The force of gravity still has not been clearly understood, and attributing a change to what must be eternal and absolute sends one in the wrong direction.

Whatever force field or wave gravity is, perhaps also

lays beyond our instruments to comprehend other than as some kind of quantum mechanical probability. What happens to the mass-less photon, considered to exhibit the properties of both a particle and a wave, could be simply owing to it's existing,being energy, even without mass. Photons are possibly being pulled by gravity, preserving the nobility of space. No evidence proves conclusively that space need change in any way, and besides, without knowing the Absolute Frame of Reference, which we will discuss more in chapter 9, we cannot tell individual identical sized sections of space one from the other anyway, so how would we know?

That being said then, a new theory of gravity perhaps that fits all that has been discussed here will be developed, and space being unchanged and eternal doesn't invalidate Einstein's theorems either,with some modification noted whereby space and time are two separate entities possibly, with only one, time, possibly admitting of variation.

A lot of the physics scientists do involve space coordinates, x and y axis type representations, and it is very convenient and expedient for equations that involve space and time to factor the two in together. I'm really not sure if that isn't what Einstein meant by space/time. It is a construct to facilitate one's understanding of the processes, but the actual reality is that space is perfectly flat and perfectly empty until something enters it, and it passively permits passage.

Computer simulations of all this might even be able to come up with a very accurate 72 hour prediction of meteorological activity when one conducts an experiment with copper, and all the other figures we came up with for the primordial specks are inputted, as magnetic monopoles. Chaos Theory says it can't be done, but after all, if the force extends far enough, and is strong enough, and if the primordial speck is unstable enough, small enough, and abundant enough, I think the computer could get it all right and give us an accurate prediction.

A computer could accurately gauge a quantity of primordial specks moving along with the atmosphere with 10 times more mass and energy than the N_2 and O_2, and also successfully estimate how much of an increase in quantity of these primordial specks would occur, given the weight and position of the copper. It would then, based upon how much mass had accumulated in the area, assign motion to the N_2 and O_2 particles based upon the gravitation of the total objects now present. Adding an increase of water molecules from 0.25% of the water molecules in the atmosphere to 3% of the atmosphere in that time to the program, based upon the likelihood water molecules will be following the path of least resistance by a certain factor depending on the height and quantity of copper, could make the computer simulation even more accurate. So all the predictions made by the theory could maybe be seen to happen in a simulation where all the conjectural data of the theory had been added.

Some of the newer theories brought forth in cosmology include the possibility of other dimensions. Multiple universes have been hypothesized to exist right along with the universe we live in; we just don't perceive these other universes because they are in a different dimension. Ten or eleven other Earths traveling along with the one we live on, where everything is almost the same as here, is the scenario that has been proposed, or at least that is what the amount of dark matter and dark energy suggests.

This multiple universe idea has gotten quite a bit of attention in the world of science fiction. A number of new movies and TV programs have emerged with multiple universes as a central theme. "Sliders" was a TV serial that involved passage between our universe and other multiple universes."Fringe" is a one hour show with one other alternate Earth in a parallel dimension. The puzzle of where the dark matter and dark energy are in our universe could be explained in this way, though it looks as though confirming the existence of such other dimensions and proving they exist may prove difficult, if not impossible.

An infinite number of other universes have been hypothesized also, though one would wonder how there could be more matter and energy than has been apparently interacting with our universe. If we took the estimate of 10 or 11 to 1 for the dark components to actually represent other dimensions then it would be stretching the imagination to consider that there were more than about 12 different dimensions, the other 11 beside our dimension being similar representations of the same universe, or perhaps anything.

The multiple dimension idea is about the only other feasible explanation for where all the missing energy and mass are besides the theory of primordial specks. As to which theory is correct, we will have to wait for more evidence. The multiple dimension theory is weaker than the primordial speck theory in the sense that other unexplained questions are explained by the primordial speck theory that are not explained by the multiple dimension theory.

The most notable difference is that the primordial speck theory explains where hydrogen originates, whereas with multiple dimensions that question is left up in the air. There is also the anecdotal evidence in chapter eight that any conscientious investigator would have to consider. The rain falling where it shouldn't is indeed the most intriguing bit of evidence, although water molecules could be responsible for most of those effects. The uneven distribution of galaxies also winds up unexplained with the assumption the known universe is everything and it has multiple dimensions.

What happens in the first moments of the Big Bang, a big question in astrophysics, was so tumultuous as to have torn asunder all the matter and energy to such an extent that all that was left was magnetic monopoles. These eventually condense into hydrogen once some paired hydrogen isotopes wander in from afar, and that process is not finished presently. Somehow neutrinos come into existence also. The known universe contains approximately one hundred billion galaxies plus ten times that in magnetic monopoles, primordial specks.

Black holes will eventually be seen as the retrofitters of the universe. The second law of thermodynamics states that entropy within a system will always increase. The universe is always moving toward a more disorderly state. The only time or place where entropy doesn't increase, one could argue, is within a black hole. Within these enigmatic entities gravitational forces prevent almost everything from escaping, and the pressure is so great that molecular bonds are broken. They are a system unto themselves, arguably discrete and self perpetuating. They exert a tremendous gravitational force upon everything else in the universe.

Every last little thing that comes within range of a black hole will be absorbed by it, and be converted into a part of the homogenous whole. A twenty billion year old light wave, incredibly faint, spread incredibly thinly, when absorbed by a black hole, changes into part of the plasma within the black hole. Entropy decreases, as a total quantity throughout the entire universe, when that happens. A forty billion year old chunk of lifeless matter, adrift in space at almost absolute zero temperature, when absorbed by a black hole, becomes part of the plasma within the black hole. Entropy decreases, as a total quantity throughout the entire universe, when that happens.

Once a black hole becomes large enough and collides with another at high speed, the two objects being the only objects of mass over a very large area, all the plasma is released in an explosion such as the Big Bang, and galaxies eventually form with energy in abundance, planets with liquid water, life, etc. Thus, repeated cycles of systems of galaxies form, eventually burn themselves out, are absorbed by black holes, two black holes collide, and the cycle repeats. As theorized, hydrogen has supplanted any other possible reproducing entity when the plasma from a Big Bang has cooled sufficiently; either some hydrogen isotopes wander in from afar, or the known universe succeeds in making a few spontaneously, these serve as templates, and eventually hydrogen is the most abundant element in the known universe, upon which all the heavier elements are based.

The question of how things happened in the early universe with respect to whether the first hydrogen atoms developed spontaneously or drifted in from afar might be explained by the behavior of the two black holes before colliding. Gravity had to have been pulling the two black holes toward each other for some time, so the two objects accelerated as they approached each other. Objects captured by the gravitational fields of each of the black holes would still be heading towards the black holes from all sides.

However, the black holes are now accelerating, so some of the things that came within range of either would fall in behind it,so each would have a tail similar to a comet's tail as it accelerates around the sun, but much larger. This tail of things following the two black holes would continue to be gravitationally attracted to the black holes even after the two collide,explode, and begin to occupy a rapidly increasing space. Therefore, once the plasma of the early known universe cools, isotopes of hydrogen pairs are,being the most abundant thing in the entire universe in all probability, already in the area occupied by the raw material of the known universe, since they would be among the things following the black holes.

Stretching the imagination to conceive that whatever raw material the known universe consists of after exploding and cooling must somehow accumulate between the two hydrogen atoms of the stable isotope, and there make a copy of what is on either side of it, just seems the most likely event; a very similar type of reproduction goes on in living things, where a cellular part emerges between two identical copies. The point of convergence between the two electrons doing figure eights around the two hydrogen nuclei pull the tiniest things in existence into the space between the two hydrogen atoms of the stable isotope, and because of the proximity on either side of identical things emitting intense beams of energy, a third hydrogen atom coalesces.

There is information transfer, the primordial specks assimilate the information and adopt the identity of a single hydrogen atom. The isotope briefly becomes deuterium, H_3, and then the extra hydrogen atom is cut loose, to pair up with a like atom. That puts the hydrogen atom itself into a kind of atom of its own, since it reproduces as pairs. The simplest atom is the only one capable of it, as stable pairs. Hydrogen is the starting point for all the other elements also, so it is a really busy little thing. The entire cycle from black holes to systems of galaxies and back isn't perfectly repeating, all the matter from the entire universe will be spread out into more systems of galaxies and black holes than we will ever be able to see, and there are no walls in between. Therefore, the matter from one system of galaxies, as it burns out and is absorbed by black holes, could wind up in a dozen or more different new systems of galaxies eventually depending on where gravity takes the various parts of the system of galaxies burning out. The entire universe is surely too large for each Big Bang to be an exact replica of the previous. Earth is now existing for the first time, not the billionth time. Many planets similar to Earth may have come and gone, but similarities, and not exact repetition, are much more likely.

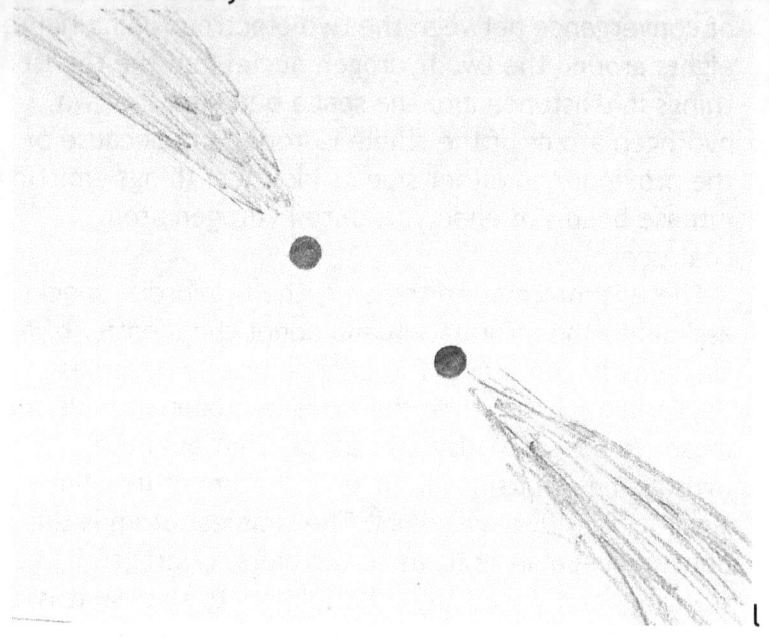

Two black holes one second before collision, matter in pursuit

Every time a big bang type collision occurs, plasma is released in a huge explosion, and the fragments remaining are completely new things, with no memory whatsoever. Therefore, happenstance will prevail in each such event. When one contrasts the assertion that no two snowflakes are exactly alike with the assertion that two big bangs could develop identically, one can see that exact repetition is extremely unlikely.In such a large system as 100 billion galaxies,size could certainly vary some between black holes that collide, and neighboring black holes to the system of galaxies could vary in distance and direction to the developing system of galaxies, changing the momentum of things in that system of galaxies uniquely in each instance. The total mass of the two black holes colliding would seldom be identical, and even if so, different gravitational factors in the vicinity would make the two events different. Considering the weight of two black holes containing enough matter and energy to generate 100 billion galaxies and still more dark matter and energy, how likely is it that wandering black holes in the depths of space would ever collide again with the exact amount of matter and energy, down to the last gram?

There could be a universal plane, just as there are galactic planes. A galaxy flattens into a disc like shape when seen from afar at the right angle. The entire universe, in other words, truly all matter and energy that exists, probably orbits around in some similar narrow disc like shape from one angle. Unfortunately, there is no place, no vantage point, that would reveal it all to an observer, since the places farthest away would always be too faint to see.

Entropy increases throughout the time as energy and matter are thrown out of a black hole, until such time as that matter and energy returns to a black hole and entropy decreases. A hot, pressurized, uniform plasma squeezed inside a small area is certainly a more orderly arrangement than objects of mass of varying sizes spread out over billions of light years and releasing energy throughout, planets with living things, etc. Things grow more and more complex as more and more interactions between various elements occurs; hardly the case when squeezed into a plasma for a trillion years. Things stay pretty much the same within a black hole, the only change would be an increase in size, or a collision with another black hole.

The centers of most if not all galaxies is thought to be a black hole. Some collisions must occur in the universe between black holes that aren't super massive enough to create billions of galaxies, given that in our known universe, there are nearly 100 billion black holes, one at the center of nearly all the galaxies.

The likelihood is that within one system of galaxies such as our own, galaxies orbit each other in clusters and when two galaxies merge, the black holes at the centers of the galaxies won't be approaching each other head on and would likely merge into a larger black hole at the center of a larger galaxy. The Milky Way and the Andromeda Galaxy are going to eventually combine, or collide, over millions of years. Black holes in our known universe after more billions of years will combine with more black holes and eventually become super massive black holes after the stars have all burned out and eventually there would only be a few left.

The Big Bang could have been more than one such cataclysmic event over the last 14 billion years or so. If, for example, a mini big bang occurs between two black holes on average every two billion years or so, a half dozen or so may have happened in our neighborhood, and if that happened within a larger universe, the gravitational pull of the rest of the universe would have all the observable matter from our perspective heading away from us at increasing speeds, just as we see now. So, there would really be no distinguishable differences between one big bang around 14 billion years ago and a half dozen of them spread out over the same number of years from our perspective.

The microwave background radiation might tell us more about that, but I don't know for sure. For instance, if a mini big bang occurred less than 3 billion years ago, the background radiation from the event would have waves that were less elongated than background radiation from much longer ago. But if all background radiation detected is all the same wavelength, that would rule out multiple big bangs unless we are only detecting the background radiation from the most recent mini big bang, and all the other previous ones are now too faint to detect at all.

100 billion galaxies combined with enough primordial specks to generate another trillion galaxies is quite a lot, so I'm guessing there have been more than one colossal explosion over the billions of years that our known universe has been around, in its current state. Maybe the most recent was 5 billion years ago, and that is all we now detect, and it was the fourth or fifth such event in the history of the known universe.

However it all happened, with dark matter and dark energy occupying such a high percentage of all that exists in the known universe one could argue that this suggests that perhaps not all matter found within the known universe was contained within one big bang, since if it were, more hydrogen would have surely evolved by now, that event having happened almost 14 billion years ago. The lack of developed hydrogen suggests that some of the known universe might have come along more recently, with less time for hydrogen to develop, and has conjoined with what was here before. The galaxies we can see are all from the first few events, and most of the dark matter and dark energy that is now intertwined with the galaxies came along more recently on the cosmic scale.

The Wilkinson Microwave Anisotropy Probe, or WMAP, is an ambitious project designed to map the entire known universe from the background microwave radiation it receives. Pictures of the map have been displayed. Comments to the effect that the probe saw all the way back to the plasma of the early known universe were made. One has to sit back and think about that. After all, the plasma of the early universe became the 100 billion galaxies or so that now exist. The Milky Way is one of those galaxies. Therefore that plasma contained the Sun and Solar System as well, and what became us.

If we shine a bright light into the heavens, for example, the light on top of the Luxor Hotel in Las Vegas, do we expect to ever see that light again? No, we would never see that light. It left Earth straight up at the speed of light never to return. How can the WMAP, then, have seen something that shone from the something which we were within when it shone? It should not be returning to the point of origin of the light. Maybe the cloudy haze the scientists saw at the edge of what could be seen originated from the Mystery Boom that NASA scientists detected back in 2006. It surely tries my grasp of the laws of physics to explain how a light that shone from somewhere could be seen 13 billion years later from that same somewhere that is now considerably bigger. The light that shone then surely outstripped us as the known universe expanded. No mirrors exist, surely, in the depths of space.

I see nothing to stop black holes in our known universe from continuing to grow, merging with other black holes and eventually becoming the only thing around for billions of light years, at which time such a super massive entity would pull, and be pulled toward, others of its kind. When that happens extremely huge black holes travel really long distances straight toward a like object, collide head on, and lo and behold, a Big Bang. What must be contained within a black hole fits exactly the material ejected in the Big Bang.

Galaxies that are observable now are each being slowly absorbed by the black hole at the respective center of each. Since super clusters of galaxies orbit around each other, eventually the black holes will combine. In the end, three or four remaining super massive black holes head off in opposite directions, being already far apart in the expanding known universe,towards other black holes from other systems of galaxies.

Or possibly, after 500 billion years or so, there are a few thousand black holes remaining to our known universe, and they all drift off toward other objects of mass elsewhere in the entire universe, to continue growing until they are so large that a significant area of space lies between them and a similar object toward which the black hole is pulled, resulting in considerable speeds and a huge impact.

Contrary to popular belief then, perpetual motion is possible, all you need is gravity. The fact that the known universe is, and is here now, confirms this. The simplest explanation for how the wilderness we call the universe exists is that it always has. No special contrivances are needed for perpetual motion to have always existed. If it is a true depiction of reality, that the universe is a system in perpetual motion, just by dint of being incredibly large, then it existing now more than likely means it existed in any past time. Removing something that large would take considerable doing; I really doubt there has ever been any agency capable of it.

The first law of thermodynamics, that energy is conserved within a system, holds at all times in the entire universe, no doubt. The entire universe probably hasn't changed an ounce in total weight ever. Time travel will prove an impossibility. The universe will march on in the inexorable present as it apparently always has, completely uncaring, huge, impartial. That makes the cells of our bodies the oldest things in existence, as everything else in the universe is also. That consideration doesn't change the fact that for those cells to remain what they are, and not change into something else, realities must be faced, hunger overcome, shelter obtained, etc.

It is still a fun thing to think about, that the universe may very well have always been in existence. One could speculate that after vast eons of time the things that make up the universe might evolve through experience to some extent, and that the big bang that we are part of now is vastly different than a big bang that occurred billions and billions of cycles of black holes and systems of galaxies ago. That question won't be answerable without retreating to the distant past, something humanity cannot do. Explosions of the magnitude of big bangs probably reduce things in the universe to such small things that retaining any information from the past is probably impossible, so it could be that the current state of the known universe is not at all unlike a system of galaxies that developed a very long time ago.Residing in a black hole under tremendous pressure for billions of years might tend to erase any residual information some parts of the black hole may have had. Could the particles themselves experience déjà vu, having had a similar experience billions of times? Could some of the inspirations experienced by humans over the years been because the particles of the human beings experiencing the inspirations have existed for all eternity?

Huge objects of mass outside our known universe but not that far away would be unlikely to be observed in an area of space that was relatively clear since these objects would have pulled things from our known universe towards them,so the space between these distant objects and us would be filled with galaxies.

If more distant objects exist, and some of those are likely to be black holes,we should eventually expect to see among the most distant galaxies that one that was there has disappeared, and then another and so on as they are absorbed by a super massive black hole.

Unfortunately we probably have a long time to wait, since the events we are witnessing from the most distant galaxies occurred many billions of years ago, and the light has only now reached us.If a black hole is absorbing a galaxy visible to our telescopes right now at the extreme edge of the universe, we won't find out about it for another 14 billion years or so. The loss of an entire galaxy wouldn't take place overnight, either, the process could take millions of years.

It could just be that the hydrogen atom in its current form has taken up such ubiquitous residence in the entire universe that any time two black holes collide and explode, hydrogen of the type we have, not antimatter, will develop, and there are no antimatter systems of galaxies about. Nice to know we have questions that we will probably never definitively answer.

As corny as it may seem, I just don't see how the universe would be able to do much else besides what it appears to, which is create huge systems of galaxies from black holes colliding. Some of the new cosmology books get real far out when it comes to explaining what is going on vast distances from us. Vast distances from us the laws of nature will operate just as here. The only way for the universe to create new energy is through black holes; when they collide and the second law of thermodynamics begins to operate, it will always result in a scenario similar to what we have in the known universe.

Hydrogen will start to come into existence, and eventually additional elements, stars, planets, life, all come along in the course of time. Extremely complex entities come into existence such as human beings,with even more fantastically complex problems, such as the subject of this book. Extrapolating from what we can observe to be occurring to the conclusion that such events are ordinary seems reasonable. Obviously, we will never know for certain what takes place in other areas of the universe beyond the known universe, or for that matter most of the known universe. That does not mean we cannot develop a keen sense of how the universe must in all probability exist in places we will never see.

The complexity of things in the universe comes from the extremely small size individual components can be. The human brain has 10 billion neurons, each one a cell with billions of atoms of hydrogen, carbon, nitrogen, oxygen and other lighter elements. Somewhere in the known universe is a waterfall hundreds of times larger than Niagara Falls. Somewhere in the known universe there is a much more spectacular mountain than Olympus Mons on Mars. To see all that must be out there would reveal all kinds of different life forms, geologic formations, an endless panorama of new and wondrous things. The same would be true of any system of galaxies that come about. It too would be basically the same, with individual differences. Our imagination is going to have to suffice to fill in what we will never see. What is difficult to imagine, is that something as large as what the entire universe appears to be, would have been found at some time in the past to have not existed at all.

The current Big Bang theory, that everything, including space and time, condensed into a singularity, or emerged from one, if true, gives us a catalyst for the explosion known as the Big Bang. Space, time, matter and energy cannot compress indefinitely. Something gives. If, however, space couldn't have changed, and not everything was condensed into the hypothetical singularity since there are other objects of mass and energy beyond the known universe, which couldn't be there without space and time, then the theory falls short in explaining the Big Bang, and is also without a catalyst for the explosion. The collision of two black holes at high speed gives us the now missing catalyst. It also gives us continuity, something that must surely be there. To suppose that prior to the big bang there was nothing, isn't extrapolating very well from known facts to reasonable or inevitable conclusions.

Now that we have the evidence from the NASA discovery in 2006 of extremely powerful radio waves, the likelihood is very good that this could prove to be a big bang in its own right. Getting back to the Grand Unified Theory of a single unified force being split apart and as a consequence enormous numbers of magnetic monopoles being created by the Big Bang, that theory would work equally well if the single unified force existed within black holes before they collide.As we've shown, everything within a black hole would be basically the same and mimics entirely what the Big Bang started as, a super hot plasma. That everything in a black hole is subject to a single unified force before it collides with another and explodes makes sense, it is all one type of thing.

A neighboring Big Bang would create magnetic monopoles just like any other, and since radio waves from the NASA Mystery Boom are already here, chances are pretty good we are now being flooded by an enormous amount of additional magnetic monopoles ready to turn into hydrogen. Until we have more evidence it would be hard to predict what that might do. The Sun could increase stellar activity as a result and we could be on our way to a swamp like Earth in just a few thousand years. More hydrogen on Earth means more water, and with higher temperatures, melting ice caps, and swamps. We have no idea how long this other thing has been there, either. Could be all the possible effects from the introduction of more hydrogen precursors to our known universe have already come and things have stabilized. Humans have just noticed it recently, but it may be billions of years old. Indeed, the discovery in 2006 should eventually point to that event being responsible for a lot of our current dark matter and dark energy.

In a Cosmological sense, the observation that the lighter elements within the core of the Earth are continually forcing their way toward the surface that is a factor in the development of volcanoes has tremendous implications when one considers whether or not it is likely that there is other life in the universe. If supernovae create heavier elements, and planets get created at the same time, the super huge chunks of molten stellar material that eventually became Earth and Mars came from a Blue Giant or other type star in a supernova event, with a yellow dwarf star, our Sun, at the center of a Solar System with numerous planets with moons, asteroids, and comets emerging at the end of it all.

Mars being smaller than Earth, and farther away from the Sun, it would have had a solid surface much more quickly than Earth, and volcanism would have brought the lighter elements to the surface just like on Earth. The scenario that gives us is Mars was probably very Earth like with liquid water on its surface maybe as far back as a few hundred million years before the first micro-organisms appeared on Earth.

Some scenarios describing the early solar system assume the planets just eventually clump together after a supernova, from dust and small fragments. Twenty miles beneath our feet, our planet is hot enough to liquefy metal. That would make it seem that the planets were more likely huge chunks of stellar material thrown out of the supernova that themselves are too small to ignite and burn as a star. The outer surface of the planet solidifies, trapping some of the lighter elements within.

Having liquid water, sulfur, carbon, nitrogen, oxygen, hydrogen, etc., and the propensity of these lighter elements to grow as ever more complex accumulations of chains of molecular compounds brings us to right about where Earth was when life first appeared. On Mars, several amino acids could have developed just as here, and with one simple entity able to reproduce itself, it began to do so,or whatever polymer made the breakthrough of making copies of itself, and eventually whatever this reproducing entity was grows further in complexity until it could be said to be a living organism. I've seen the statement that amino acids are actually small engines in themselves, and each would have different mechanisms that a living thing with various parts could put to use.

Recently a program aired about the search for the origins of life, the DNA molecule, and the genetic code, and the conclusion reached was that the whole thing could be replicated and would turn out to be pretty simple,but one would need to know what the very first steps were. What original polymer gained complexity and became living, and how, is the question.

Naturally occurring phospholipid molecules have two ends, one hydrophobic, the other hydrophilic. What happens when these molecules are placed in water is they naturally arrange themselves in a membranous ball, with the water seeking ends of the molecules facing outwards and the water avoiding ends aligned to the inside of the ball. Cellular life must have taken advantage of this naturally occurring phenomenon since cell membranes are still similarly designed.

Multi-cellular life might not have been possible without phospholipid molecules. Some polymer begins to make copies of itself. One of these is enveloped by phospholipid molecules. The polymer reproduces again, and the phospholipids split into two balls, one enveloping each of the now two polymers. Populations of this thing, not quite life yet, begin to develop after time. Some wind up separated from others of its kind by some distance and encounter different environments. Subsequent generations develop differently, depending on where they happen to be. At what point the polymers surrounded by phospholipids are decidedly living might be difficult to distinguish. The entity absorbs molecules in the sea, eventually incorporating amino acids into its structure as it grows more complex.

Life developing on planets orbiting stars among the 100 billion galaxies of the known universe, when one considers that other planets will have volcanic activity and lighter elements spewing out all over the surface some time after they are created in supernovae, would have to be inevitable. The second law of thermodynamics and the extremely small size of individual atoms combines to create increasing complexity among the lighter elements. These lead to living things, and in the entire universe that would have happened many times.

Plant life, marine life, terrestrial life, and avian life would be found as life flourished in every available niche, employing differing survival adaptations. Plants on other worlds would have appendages holding them stationary on land and underwater, and similarities here, too, could be inferred, roots and branches are the appendages of plants, and plant life elsewhere would also branch out in some way or another for nutrients. Marine life would develop a means of extracting oxygen from water, flying creatures would develop ultra light materials for wings and feathers, for the most part. Maybe in some instances gravity is slightly stronger than here, and avian life fails, not for lack of trying, but simply because things can't fly fast enough, or long enough, to evade capture by predators. At what specific gravity does avian life face inevitable failure as life develops on alien planets with liquid water, amino acids, etc., would be an intriguing question to answer. Somewhere are planets suitable for some living things but too heavy for avian life; it is unknown whether humans ever visit any planets and see such things in the far distant future.

Other assumptions can be made about living things on other planets. If the environment is a safe enough haven for multi-cellular life to take hold, it probably will. Once that has begun, survival of the fittest will leave creatures that survive only. Since the development of sense organs, central nervous system and appendages developed here as the battle for survival among animals with movement raged, one would have to assume that those three attributes would probably be fairly common among creatures with some form of movement, on other planets.

A head and eyes and ears, then, could turn out to be a lot more commonplace than one might think. A mouth with taste buds, some kind of olfactory organ, all these kinds of developments could be predicated from the need to adapt and evolve, just like here.

The same things that course through our atmosphere, course through alien atmospheres. Light waves in various spectrums, sound waves of varying lengths and intensity, odors of numerous things, would be as abundant there as here. Eyes, ears,and nose had better get developed or said creature goes extinct. Eventually some creature succeeds and reproduces. After a million years numerous different species are found, all with the common ancestor that successfully developed the sense organs to survive at a much more primitive level.

In this hypothetical alien world it is now difficult to find an animal that does not have the survival equipment of various sense organs, a brain and nervous system, and four appendages. Quite possibly some 6 legged and 8 legged creatures will develop, and just like here, be smaller sized creatures than the 4 legged ones.As here on Earth four legged creatures in the alien world would have, on the two appendages nearest the brain, specialized adaptations to help the creature in possession of them to capture prey, break open seeds, and other manipulative activity. Eventually one of the more adept would gain the position of apex creature. Plants on other planets would in some cases produce fruit that are edible for animals but also propagate the plant species, and probably poisonous fruit would develop among some species. The DNA molecule probably has some very close relatives. All the tiny chemical changes going on, things like photosynthesis, oxidation, are as equally likely on far distant planets.

Here on Earth, there is around 21% oxygen. Why this should be so wasn't clear until not too many years ago. If the percentage of oxygen in the atmosphere rises, fire burns more readily. The laws of nature dictate how readily fires burn, and for that curious reason the Earth cannot add measurably to the percentage of oxygen in the atmosphere. Were an excess of oxygen to begin to occur, forest fires would consume the excess oxygen and the level in the atmosphere would fall back to 21%. That same type of constraint on the levels of oxygen would be found on other worlds as well.

Naturally we are curious about life elsewhere in the universe, actually meeting other intelligent beings is another thing. The distances to other stars are huge, and the time involved in travel long and dangerous. It could happen, but even if it doesn't,we could probably rest easy in the realization that life would have gotten started on other planets as well, some of these would have eventually led to intelligent life, and so on.

Chances are an intelligent life form developed quite a long time ago. Recent estimates based on stellar evolution, and relative abundances of metals in stars far distant conclude that life on planets such as ours probably didn't become possible until the known universe had reached 10 billion years after the Big Bang. One could hypothesize ancient aliens capable of moving from one system of galaxies to another, and though unlikely, a civilization trillions of years old is theoretically possible.

If we are ever visited by aliens,it will be by exploring, curious beings that would attempt to communicate. They would have developed culture and explored the natural sciences, just like us, but for a far longer time. Technology to translate alien languages might eventually be possible, just based on how fast computer processors are doubling in speed in our civilization. Expect aliens to understand us, then, if they made it this far, and we can expect to understand them if and when they communicate because of greater computer processing abilities in the not too distant future. At the very least, the aliens would be able to say something using one of our languages in printed word form, possibly even have a machine that could make the sounds necessary to communicate with speech.

The aliens would have the advantage of being able to study our language longer, as they approached our planet, than we would have been able to study theirs, if an alien species ever made contact. Likely the aliens would send radio transmissions in one of the major languages of Earth after listening to us for a while. We wouldn't need to know their language, though it would eventually be translated if we became friends with the alien species. I like to think it possible, and maybe even happen one day that we meet alien travelers.Truly happen. Any contact by intelligent species would be minimal because of possible contamination. They would be wary of corrupting life support on the vessel they travel in. Just opening a spacecraft door on Earth could bring in enough molds and spores to completely overcome the living things within, depending on what eats what, something not easy to answer until it happens.

No contact if a species doesn't control its environment

Collecting specimens from Earth would be done carefully and clandestinely, to lower the risk of problems. They would not involve us since that would further complicate a simple extraction that would not harm us materially. Any contact between life forms on one planet and life forms from another planet could have any number of consequences, most of which we would be unable to predict. Molds and spores are just the first on a long list. Foreign DNA ingested by terrestrial animals or plants could begin changes we could only guess at. Viruses that are dormant within visiting beings could run rampant within us, or vice versa. Intestinal bacteria could switch species. Different enzymes from food from another planet might start chemical processes difficult to predict. We have to assume that most life would be very similar to life on Earth, but small differences on the scale of cells and DNA could still make food from another planet risky. Of all of the dangers of space travel, the inability to find food on other planets that is safe to eat may be the one that dooms most alien space travel. With a spacecraft of considerable size,a section could be put aside for the growth of food. That adds more moving parts and other complications to the trip, increasing the likelihood of a malfunction becoming life threatening. With no success at finding anything to eat on alien planets, the things grown on the ship would have to sustain the alien travelers. A space craft with no survivors could drift by, the travelers having starved to death a few million years before.

An alien species, if visiting Earth might take samples quietly, do some testing on plant and animal matter, and decide if the organisms on Earth are compatible with them. If they were to decide that Earth was a habitable biosphere for them, they might then initiate contact. The fact that this has not happened indicates we have never been visited,or in the very unlikely event that a visit had happened, the aliens decided the life forms here were not compatible with them, and left.

Curious that if intelligent species developed elsewhere in the universe, and this most likely has happened, they would have developed on a planet with liquid water, in all probability, as we have, and liquid water on the surface there would have evaporated and joined the air, just as here, and one could say that these intelligent aliens also experience weather.

They, too, would have also noticed that copper, or anything that conducts electricity well, and lead or some other non-conducting material in pure form of some specific quantity in an elevated location can cause changes in barometric pressure,and precipitation amounts, at some time in their development as intelligent beings. There could have been difficulties assimilating the new information in other species, as well. Weather could prove to be the most discussed topic in the entire known universe, not just Earth. It would be hard to take a survey, though.

Chapter 4.Weather Modification

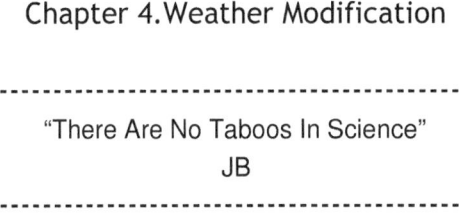

"There Are No Taboos In Science"
JB

Nikola Tesla had ideas about providing abundant water to arid regions, controlling flooding and other weather control ideas. These ideas, and many others, were never developed by him.Those familiar with Tesla know that he was a prodigious inventor with numerous inventions and patents, and he was usually in financial trouble. A method of modifying the weather would not likely have yielded a weekly paycheck for him, whereas a number of other ideas he had offered a better chance of financial returns. Not to say that Tesla himself thought it through like that, more likely he just pursued various things as he saw fit,such as building an alternating current motor. Whatever struck him as he pondered his ideas is probably lost to us for all time, what we do have is all his unfinished business.

Tesla tore up his contract with George Westinghouse, throwing away his right to royalties on his alternating current invention, which paved the way for Westinghouse Electric to be able to afford to switch over to alternating current from the direct current in use at that time,championed by Edison, but far inferior to alternating current. Whatever his motivations and decisions were,we are still blessed by many things that Nikola Tesla invented and developed. Maybe we should add one more, and move on with this discovery. It has been claimed that an agency of the U.S. Government seized Tesla's private papers upon his demise.

Charles Hatfield came along in the early part of the 20th century, sometime between 1900 and 1920, and was known to have claimed to be able to make it rain. A meteorology text where I learned of this man stated that he was paid by the mayors of a number of towns and cities to work his mysterious cure for drought stricken areas. The book also mentions that he seemed to be successful in his attempts. The places where he was hired to do this work actually did see rain soon after he was hired and performed his duty.

What Charles Hatfield did was burn some copper compounds in a pyre type furnace, copper sulfide or sulfate, releasing plumes of copper compounds that rose into the atmosphere and joined the gases and dust that was already there and began to follow the prevailing westerlies just as the rest of the gases and dust do. He also claimed a number of different ingredients went into his concoction that he burned. Apparently Hatfield himself was not anxious to reveal exactly what he was doing.

The author of the meteorological text I had perused regarded this Charles Hatfield as a kind of charlatan. In the opinion of the author these copper compounds or whatever Hatfield sent airborne didn't really have any effect on the weather. He went on to state that there have been similar claims about the presence of some metal or another having an effect on the weather, and concluded that these were all just instances of wishful thinking, not anything with a basis in fact.

That was that, apparently, all in one paragraph in a

book with a part of the title "Weather Modification", copyright 1980, some 400 pages in length. I never read the entire book, but I find myself wondering what the author could have possibly filled it with, since the only thing that changes the weather, according to meteorologists, is cloud seeding, and that only a little, and unreliably at that. Why, indeed,write a book about a subject that, according to the author and his peers, has so little subject matter? I looked up the title since I couldn't remember the author's name and found three books that could have been the one that I skimmed through and found that one paragraph and the mention of Blue Nile somewhere else in the book. These three possibilities are listed in the references at the end of the book, all copyright 1980.[7]

   That is apparently how things currently stand with regard to the subject of this book. Whereas with the first edition the possibility that the water molecule could be involved was overlooked, here the author is fully cognizant that the evidence is currently overwhelmingly on his side once the polar nature of the water molecule is added to help explain what in the atmosphere is moving where during such experiments as previously described. The three authors must have known the water molecule would be likely, as water vapor, to respond to an external electrical field such as would be created by a row of copper tubing in a high location. Why, then, does one of the authors state otherwise?

   Blue Nile, or Nile Blue, is a military research program classified in 1945, and according to the author of whichever book I had read, was still classified as of

1980. Recently a program aired about military weather modification where the trail in Vietnam used by the North to infiltrate the South was in the 1960's cloud seeded by the U.S. military, and this activity appeared to flood out the trail as often as not,and this was what Blue Nile was about, according to the presentation. For thirty years Nile Blue intrigued me, because it was classified, and because of the subject of this book.

What little I know of chem trails, as they are called, is that some additive is added to jet fuel to make the exhaust different, or something is added to the exhaust as it exits the engine and depending on what those chemicals are,this kind of activity could prove effective in creating a path of least resistance. The similarity to this type of weather modification activity and the type of thing Charles Hatfield did are apparent. Hatfield, on the ground, maybe in a high location, introduced chemical compounds to the air in a furnace, with the rising heated air ferrying the compounds upwards to the height of clouds. The height of the jet that is releasing tainted exhaust could vary, but flying through the air at any height is a pretty good release point for a chemical that could prove effective in modifying the weather. In comparison, chem-trails seem more direct and perhaps more effective.

With both ground based and air delivered exhausts, whether from a furnace or a jet, a fantastic number of tiny particulates are released airborne that water molecules can latch on to and begin accumulating into raindrops, and if the molecules happen to also create a path of least resistance, there is the certainty that barometric pressure will fall, and clouds will begin developing. The developing clouds and the particulates ready to start raindrops might work together well.

However, one would have to consider the cost of being without whatever chemicals are burned each time since once they become exhaust the chances of recovering the material is non-existent. Copper placements could always be adjusted in amount to accommodate the creation of storms of the desired size, and, given safeguards against theft, always remain in the possession of the weather modifier.

The earliest smelters of metals could have even picked up on this back in the bronze age. It would be interesting to see if some translations of ancient writings may have been interpreted wrong and actually refer to precipitation and metals. There was extensive irrigation in Roman and Mayan cities, could it be they knew they could depend on rain happening? The early smelters of metals had not developed a technique for separating copper from zinc. What they wound up with was a combination of the two metals, bronze. Bronze is also a good conductor of electricity, probably only slightly less conductive than pure copper. By the time of the Roman Empire copper may have been purified by chemical means, the Mayans maybe not.

Suppose the ancient smelters of bronze had an open pit mine some ways up a hillside. They find a vein of metal ore, hack at it and shovel it out. They dig a pit on the spot, to burn the ore and purify the metal. Some ancient settlements have evidence that this was the typical procedure. They had no bulldozers, no trucks, why move the ore? They worked it right near the mine they hacked it out of.

After a day of smelting a weight of bronze ingots accumulates off to the side of the fire pit. After about the fourth day, once the fire started, it started raining, so the fire went out. The crew then began to transport the bronze ingots down the hillside to the town. Finishing that, they waited a day for the rain to stop. Then they climbed back up the hill and after a couple days of digging, got the fire started again and began to produce more bronze ingots. Fourth day into the smelting with the fire going, it starts raining again. Ingots piling up on the side of the fire pit. This happens almost a whole summer until one time when it started raining they didn't bring the ingots down the hill because there was some festivities in the town under the big tent. This time the rain doesn't stop, but continues for almost three days.

The crew leader after the third day sends the crew up the muddy hillside to retrieve the ingots. After the rain stops the crew goes back up the hill and starts digging out more ore and eventually get the fire going again. The crew leader brings a couple young boys from town along once the fire gets going, and they transport the ingots down the hill to the town as soon as they get a cart full. It doesn't rain on the fourth day this time, and the crew leader isn't surprised. He explains to his boss that the bronze on the hillside appeared to make it rain, it usually took about 72 hours after some ingots had accumulated on the hillside and the ruling class of the Roman empire finds out.

A not impossible scenario such as that could have happened in the Roman empire and in South America with the Mayans. Probably in both instances, only nobility would have learned of this, had the process been noticed back then. Those giving directions on how to build irrigation canals wouldn't have shared the insight with lesser citizens, and with the fall of those civilizations this knowledge would have been long forgotten, those once aware of it long dead. Bronze or copper, it wouldn't have mattered. What mattered is did anyone notice that the ingots piling up made a difference in frequency of precipitation.

The atmosphere is comprised of vast numbers of sub-microscopic gases which have been fairly well described; about 78% nitrogen, 21% oxygen, a little argon and tiny amounts of carbon dioxide,methane and a few other gases. There is also water vapor in varying amounts, static electricity, sunlight, and dust, and possible primordial specks as before mentioned. All of the particles mentioned, not including static electricity or sunlight, are very small satellites in orbit around the Earth in the same way as the Moon is a satellite in orbit around the Earth.

The difference in size is quite pronounced, however, and that difference results in the orbits of all these very tiny satellites being very easily perturbed by all the neighboring satellites. These satellites that comprise the atmosphere are also much closer to the Earth than the Moon, so the uneven terrain of the Earth also factors into how easily and frequently these satellites are perturbed in their orbits.

Laws of inertia hold that an object in motion tends to stay in motion. This holds true even for the tiniest things. The fact that the orbits of these atmospheric objects are so easily altered doesn't change the fact that in the absence of anything causing them to alter course these objects would proceed in an orbit around the Earth similar to that of the Moon. The ease by which they can be made to alter course, then, is evident.

The magnetic field of the Earth varies at different locations around the globe such as the north magnetic pole and this can alter the orbits of our atmospheric satellites. The Moon is responsible, along with the Sun, for creating ocean tides, and the changing tides also influences some of the atmosphere's constituents. The Jet Stream has a huge impact on the direction of flow of the atmosphere.

Just trying to change the weather

Every moving object on land, or sea, or in the air changes some of the orbits of some of the atmosphere. A whale breaches off the coast of Alaska, a jet plane flies form Los Angeles to New York, I wave my hand, all these movements of different things send atmospheric constituents scattering in various directions. As the Earth spins on it's axis, which is what gives us the 24 hour periods of night and day, atmospheric satellites are pulled along with the rotating Earth.

Hence we have winds that are known as the prevailing westerlies. These winds blow from west to east almost everywhere around the planet. At or near the poles these winds are not as clearly defined because the top and bottom of our spinning planet doesn't travel as far in a day's revolution as the land areas away from the poles. Standing at the North Pole every direction is south, so at that precise location, there are no prevailing westerlies! Equally true of the South Pole, every direction is north from there.

Currently it is widely supposed that gravitational and electromagnetic forces are a wave phenomenon. All matter exerts gravitational and electromagnetic forces. These are thought to be aspects of the single wave that all matter propagates at the speed of light equally in all directions. These forces weaken with distance. Gravity is far weaker than electromagnetism, as we mentioned earlier. The gravitational field of the Earth is strong because the Earth is a huge object, and a man made flying craft must expend considerable energy to escape this field.

Copper is an element well known for its ability to conduct electricity. The reason that it conducts electricity so well is that copper atoms together in a piece of copper naturally arrange themselves in orderly rows. When an electrical current encounters a group of copper atoms, it passes between the atoms at the speed of light,encountering little friction in its passage. As with all matter,copper also propagates gravitational and electromagnetic waves at the speed of light.

Tiny air molecules, static electricity, dust, primordial specks, water molecules and whatever else is airborne in the vicinity of a quantity of copper will encounter the neat, orderly arrangement that copper exhibits. All of these airborne objects would encounter less friction near the quantity of copper. A path of least resistance would exist in the vicinity of this quantity of copper, a path which tiny, free floating entities would flow more easily along than in any other direction.

Water molecules will align themselves along the path of least resistance. Copper as stable isotopes has a net positive charge. With a positive charge to the copper tubing, the negatively charged oxygen in each water molecule would align toward the copper, and begin to move in that direction. Water molecules from hundreds of miles away could begin herding to the location, depending on height and weight of the copper placement.

This no doubt explains why Charles Hatfield appeared to have been successful in his efforts, and why Tesla's experiments saw intense thunderstorms. The copper compounds sent airborne, the copper Tesla coils on the platform in Colorado Springs, must each have created a path of least resistance along which water molecules and primordial specks began to flow.The electrically inert atmospheric components are gravitationally pulled along with the water molecules and primordial specks. Once the process began, the barometric pressure would begin to fall as a result of all the objects in the air traveling in the same general direction, encountering each other less frequently since two or more things traveling in the same direction will not meet.

That is the essence of the matter, the fact that two or more things traveling in the same direction do not meet. Barometric pressure is a measure of the degree to which objects in the air are encountering each other. A high barometric pressure reading indicates that the atmosphere at that point in time is undergoing more chaotic behavior, with interactions between air molecules happening frequently. A lower barometric pressure reading indicates that the air molecules are "schooling", moving in unison like groups of fish swimming together, large numbers of atmospheric constituents traveling together in the same direction and interacting between themselves less frequently.

I think most people realize that the atoms of things don't really touch each other; they come very close, but the field surrounding the nucleus, created by the electron, or electrons, is strong enough to preclude anything from penetrating within it, with the possible exception being primordial specks. A bullet fired into a tree trunk blasted into the wood, and penetrated several inches, but not a single atom touched another one. The individual atoms remained sealed within that protective boundary provided by the tightly orbiting electron(s). I guess one could say that atoms touch during nuclear fusion. During nuclear fusion in its simplest form, when two hydrogen atom nuclei come together, and become one nucleus, one hydrogen nucleus had to get past the electron of the other. The higher up the table of elements one goes, the heavier the nucleus of the element. Nuclear fusion must involve contact for the nuclei to grow larger.

Gases are quite different from solid objects.

Atmospheric gases deflect off each other quite a lot as free floating gases of the atmosphere, but don't really touch. How much that is happening is where barometric pressure measurably rises or falls. The hotter gases are, the more excited they become, raising air pressure. Land heats up far faster than water, so the faster heating also contributes to raising air pressure on land by day.

The dark matter and energy would be charged and definitely school as a result of a conductive metal of sufficient size being in a high location in an area, and the detectable N_2 and O_2 would have little choice but to follow a stream of things traveling in one specific direction many times more massive than itself.

How can I be sure the dark matter and dark energy are charged particles? Because the known particles that are electrically inert have gone through a number of changes, first becoming hydrogen, then helium, then eventually nitrogen or oxygen, then pairing up with a like atom before they reach the condition of being electrically inert. Single oxygen atoms are a charged entity, but they do not remain single for long because oxygen reacts readily with a lot of other atoms, including other single oxygen atoms.

The unknown particles are likely some fragment looking to unite with others of its kind, thus they are unlikely to be so complete as to be inert and not influenced by the electromagnetic force of the copper in the hypothetical experiment. As pointed out earlier, magnetic monopoles of such vanishing smallness as these primordial specks must be would satisfy the theory. We also have to wonder whether the smallness of whatever dark matter and dark energy are wouldn't in and of itself be enough to make them susceptible to the electromagnetic force of the copper.

Since burning copper compounds results in a loss of scarce material, and since the same effect could be expected with placement of a quantity of copper, probably 500 to 1000lbs., in as high a location as possible, that would be the most economical means to apply, one that could also be retrieved and used over and over without loss of material. Some variation is to be expected at different locations around the world since some places are closer to an ocean than others, and some locations have more uneven terrain than other places. Unusual land features might also change things.

Precipitation will usually occur around seventy two hours after the placement of the copper in a high location, with a margin of error of 8 hours. Close to an ocean to the west would probably reduce the time and quantity sufficient considerably, while in the middle of the Gobi Desert the elapsed time before precipitation occurred could prove to be hours longer, and require twice as much copper. Quantity sufficient to see the desired amounts of precipitation could vary consider-ably around the world, but even the Atacama desert in Chile on a high plateau could see precipitation with the appropriate quantity of copper positioned strategically. In that particular location, 1200 pounds of copper might be needed.

This desert would be a great place for a team of scientists to confirm the hypothesis, since at the top of that plateau it almost never rains. If an experiment was tried there and rain occurred approximately 72 hours later the idea of coincidence could be ruled out. Especially if, once that happened and the experimenters packed up and left and the desert returned to a dry condition, and then upon returning one month later, attempting the same experiment, precipitation occurred again around 72 hours later.

The surface area of hollow copper tubing being greater than solid bars of copper,hollow circular tubing would produce a more dramatic effect than bars of solid copper. If the copper is on a horizontal plane, a fairly robust wind will develop; if placed on an upward slope or vertically, there is less wind and more cumulonimbus type cloud development, and precipitation is more robust.

The idea is to take coils of copper tubing, perhaps two feet in diameter, and arrange those in six foot tall circular things resembling a bird cage without a top or bottom. The diameter of the tubing itself should be one half to one inch, though a number of various dimension tubing could be used.A number of six foot long straight bars of copper tubing could be used as supports, tied to the circular tubing with copper wire. Stand the six foot tall copper coil arrangements on end so they stand straight up, and since this will require a half dozen or more of these contraptions, place them in a row from the west to the east,with a slight curve towards north, to assist in generating counter clockwise wind flow.

If you happen to be in the southern hemisphere, low pressure systems spin clockwise down there. Therefore, a row of copper coil arrangements in the southern hemisphere would probably work best if it began from the west and proceeded to the east curving slightly south. In either hemisphere, an upward slope on the western side of a mountain or hill would work best. After about two days the sky will be pretty much cloud covered in its entirety. From the second day on, precipitation could occur at anytime, depending on when night falls, but invariably will occur by the end of three days, perhaps 8 hours longer. If the copper is removed, the effect remains for a few days to a week with lessening intensity, the effect of inertia.

Considering what would happen if one just brought an 800 pound block of solid copper and put it on a hilltop, I really don't know what would happen.I would assume that occupying a smaller space might be likely to lessen the impact it might have. It would definitely have an impact, and the density of the material might even make this type of weather modification more intense, for all I know. Until someone conducts an experiment with a solid block of copper instead of copper coils and presents the results to the public, the actual difference is unknown. Solid blocks would be much more difficult to transport.

The simple arrangement of circular coils standing on end and spread out ten feet apart as described can be accomplished by just about anyone, or several people, almost anywhere.Some of the terrain on Earth is harder to traverse than other places, especially near the tops of mountains. If one could drive a truck to 7000 feet above sea level, on the western side of a 9000 feet above sea level mountain, chances are that parking the truck and deploying the copper there would do the job. No need to begin hauling the material any further than one needs to position it effectively. Find a suitable upward slope and create a row of the things, from west to east, sloping upwards and curving slightly north in the northern hemisphere, and curving slightly south in the southern hemisphere.

Often I have seen the observation made that rules that apply to ordinary things, or other disciplines in the natural sciences besides meteorology, don't apply to meteorology because the atmosphere is too complex on the Earth. Chaos theory applies. It is true that it is a vast system with a number of unique land and water features; The Grand Canyon, The Dead Sea, huge deserts, rainforests, and oceans, to name a few. The number of components in the atmosphere is also staggering. No two rain storms are exactly alike.

These considerations, however, are no cloak for meteorologists to hide behind in a concerted effort to avoid confronting this issue. If a predictable quantity of rainfall occurs when a quantity of copper is experimented with as I've described, and if that quantity of copper being redeployed does essentially the same thing a second time it doesn't matter that there were differences in the rainfall totals between adjacent counties in the area, or that the center of the storm front was 20 miles further north than the previous experiment, or that one storm front moved due east, and the next one took a slightly more northeasterly direction.

The results would still be within the parameters of the theory, and definitive conclusions could be made, allowing for some variations between individual storms. Suppose an experiment with copper is tried when the current forecast calls for no clouds for four days with a few high clouds expected to roll in on the fifth day. A pair of helicopters sweeps the area in advance for 500 miles in every direction for unusual barometric readings, declares the area clear, and the copper is placed in its location. A result consistent with the theory occurs, for example, rain began around 9 P.M., 68 hours after the start of the experiment.

A week later a location is found 600 miles distant with a forecast for cloudless skies for a week. The crew choppers to the new site, which is swept barometrically after transporting crew and copper, and within 6 hours of learning of the forecast have the new experiment started. 71 hours later precipitation begins, a similar sized storm system of the first one. This continues, how many times? What are the rules to establish the truth of such a matter? Is that the decision of a panel of scientists?

Chaos theory predicts that if a system of things exceeds a certain level of complexity, predictions about expected events within that system are impossible. Yet if we can reliably predict the outcome of atmospheric events 72 hours into the future, this prediction holding true within certain parameters in every instance the experiment is tried, provided that is what eventually happens, we have rather proven otherwise, and have also proven that a cause and effect relationship does exist between a placement of copper on a mountain side or other high location, and an atmospheric event, precipitation, occurring around 72 hours after the start of the experiment. Once the essential mechanism has been isolated,the specific quantities of copper required at various locations could be ascertained.

Clouds block sunlight. Additional precipitation on land means increased clouds over land. These clouds would be much more effective at helping reduce worldwide temperatures than the cloud generation over the oceans proposed by the Discovery Channel Project Earth Program. The generation of additional clouds over land where things heat up far faster than water, things like concrete and asphalt, would be much more effective than cloud development out over the oceans where the additional clouds could contribute to hurricane development, especially since humanity has not yet begun to implement this proposed solution. For that to happen the process would need to be entered into encyclopedia.

Where an accumulation of copper tubing are resting can be important. A mountaintop, one would think, would be the best location. Placing the copper a hundred feet or so from the summit of a mountain, on the west side, could prove much more effective in helping a storm front to develop. The difference in barometric pressure between the west side of the mountain where the copper rests and the east side which is effectively blocked from the electromagnetic transmission of the copper is perhaps pronounced enough to create a front line between two air masses. A rooftop of a building with an additional tower added so that the roof of the entire building now has two levels could produce a similar effect if the taller wall of the new addition were to the east.

The question arises as to what happens to places on Earth that ordinarily receive ample amounts of rain when drier places begin to take in larger amounts of water. Will the Amazon or Indonesia or the Hawaiian Islands begin to see less rainfall? The answer is yes, the likelihood that some places could see less precipitation than normal is quite good, if weather modification activities began in earnest. If the residents of these areas are aware of what is taking place and are aware that they are free to remedy the situation they are experiencing for themselves no real problem exists. There appears to be more than enough water vapor available to satisfy every acre of land on the planet.

The oceans themselves would see less precipitation than normal if much more water vapor than usual were diverted toward land, so island chains would be affected. If these same residents are unaware that the weather is being modified in some areas of the world and are also unaware that they, too, could modify the weather, then there is a problem. When all residents of Earth can learn these things in an encyclopedia, there would be little doubt that should the need arise, appropriate steps could be taken to increase or decrease precipitation as needed, on a local basis.

It never rains here anymore. Did another volcano erupt?

Returning to the hypothesis that precursors to hydrogen still exist and are still churning out hydrogen atoms albeit at a slower rate than billions of years ago, do the activities of humans modifying the weather on our planet such as has been suggested here create more water than would ordinarily be created and could we be turning the Earth into a Water World, like the movie of a few years ago?

The answer to that question is probably no, since the hydrogen would be coming into existence anyway around the globe, and where a human act of weather modification was causing more hydrogen to be created at the same time in the oceans hydrogen synthesis would slow down as a result of the precursors to hydrogen having gone elsewhere, namely to follow the path of least resistance created by the human weather modification activities. The total hydrogen coming into existence on the whole Earth probably wouldn't change, so the total amount of water on the Earth would be unaffected. The Earth loses some of the atmosphere to outer space all the time, including water vapor,and the total quantity appears to be fairly stable. Here we could have the main reason global warming is becoming a problem. If enough hydrogen comes into existence daily to combine with oxygen to make a ton of water, perhaps the Earth is gaining some water daily. Measuring the total quantity of water molecules on the Earth might be very difficult. As noted, two tons of water is only 500 gallons, a drop in the bucket of all the oceans.

Maybe after a very long time the Earth will have more water but if that happens it probably would have happened with or without human activity, and be a result of continued hydrogen production in the universe that we as weather modifying beings haven't changed, except to increase such activity in some areas while decreasing the same activity in other areas on the planet.

Crop yields would see an enormous boost from the lack of severe weather and the abundance of fresh water throughout the Earth. Certain plants prefer certain environments. The litchi tree requires a place with as little wind as possible since the litchi fruit are so easily dislodged by the wind, and such conditions could be made to happen for some areas. Rainfall amounts could be increased in spring and summer, and reduced in the fall for harvesting purposes. How many times has it happened that crops have matured and been destroyed by unwanted precipitation in September or October? This type of misfortune can be averted. Any plant humans grow as a crop could have optimal growing seasons.

Thunder and lightning are usually more prevalent and likely to strike the ground when a quantity of copper tubing is placed in a vertical position on a mountaintop or building roof. The lightning could be captured by some copper scaffolding arranged so that when a lightning strike occurred, the electricity would travel along the scaffolding down to an underground circle of super conducting material of considerable size where it could circle indefinitely. Hopefully, room temperature super conducting materials will be developed in the near future. Super conductivity has been the subject of research for some time.

A few controls, some switches, and the power is

transferred to the national electrical grid. The scaffolding would have to be no bigger than would cause medium sized storms in an area. An area that tolerates lots of rain well would be needed. A crew with a cherry picker to rebuild the scaffolding would be needed after damage from lightning strikes. A ship with the entire scaffolding and super conducting ring in the hold sturdy enough to withstand intense storms could prowl the ocean, or a rig similar to an oil rig could be constructed in a remote location in the ocean, far from frequently used shipping lanes.

If such an attempt to capture electricity is made, there is a possibility that if the copper scaffolding is large enough it may even be able to channel and collect ambient static electricity in the surrounding air and generate some power without lightning striking the device, though this would be a lot less than three million volts, which is the average voltage of a lightning blast. But there are small and big lightning bolts, so some probably exceed 5 million volts while smaller ones are only around a million volts or less.

How efficient would a scaffolding in need of repair from lightning strikes from storm to storm be? What percentage of a 3 million volt blast would get captured by the storage ring, and ultimately sent to the power grid? Questions I am unable to answer. I think it would be interesting to find out, but whether or not that will ever be done remains to be seen. Once in place the only maintenance necessary would be the repair to lightning damaged parts. A design consisting of identical sized replaceable parts could be used, and the damaged parts recycled.

Some insulation would be required for the underground part, or part of the hold of an oil rig type sort of experiment in remote seas, and the crew would need a secure location for at sea and on ground activities like this. Electricity captured out over the high seas would require an electrical cable along the floor of the sea, assuming humanity doesn't decide to transmit power around the globe as Tesla envisioned, and the impact on local climate would need to factor in with any land based such experiments.

What is clearly possible is that through examining this presentation eventually by enough people, and after some time for civilization to digest the new information, and new super conducting materials in production, perhaps a group of investors with a team of physicists and engineers might actually have a go at it.

Computer simulations with the new information added would probably give an engineer or physicist a good prediction of events surrounding a rig all alone in the South Pacific with a significant quantity of almost perfectly conducting materials in the hold and also part of the mast and riggings above the deck, or the climatic changes expected with a try at it on land.

A small arrangement in keeping with the size required for a local region affected by such an arrangement to accumulate the desired quantities of water could be engineered. Additionally, this same quantity of copper could be connected to a storage ring should lightning strike, and the likelihood of that would be pretty good, considering the conductivity of the copper and its height. A two hundred square mile area has weather plans and a source of some electricity.

Near the top of the highest available escarpment, a cave is excavated, or an existing cave modified, and a storage ring housed within. At the entrance, two sided panels are mounted on wheels to serve as a door; one side has a coat of lead, and that side is exposed to the atmosphere when the rain making device is inactive, and the other side is coated in copper. When rain is desired, the doors slide open, exposing the copper side to the air, while the Tesla Coil type arrangement is wheeled out onto the hillside.

After the desired precipitation amounts are obtained, the rain making coil is wheeled back in, the doors are closed, and the copper no longer is exposed, but rather the lead exposure begins to raise the air pressure, or at the least, block the electromagnetic waves of the copper. The coils suffered several lightning strikes, and are repaired in the safety of the cave, behind the lead shield.

The electricity from the lightning helps the world convert more quickly to electric cars. Worldwide, four or five hundred such devices could fit comfortably, perhaps more. Naturally the whole process of putting things like that to use would take a long time. Every island chain in the oceans would eventually have one. Pitcairn's Island would have one of its own, being a ways away from any other islands. The electricity probably wouldn't be as significant as the rainfall, but some would be better than none. The only maintenance necessary would be replacing lightning damaged identical parts.

A steady condition of a sufficient quantity of copper being left in place indefinitely in the highest available location will yield a storm front every 3 to 4 days. Once the storm passes through a cold front with winds from the opposite direction of the storm will blow for a day or so followed by renewed cloud accumulation and the winds swinging back around to originate in the southwest, or northwest, depending on which hemisphere the experiment is in. This cold front that follows a three day event, after some hours of rain, is a very important cooling factor. The more times a placement of copper tubing produces a storm front over land, one can also count on the cooling effect of the cold front that follows, and all the clouds that develop that are themselves cooling influences over land during the daytime.

Leaving it in place may eventually see a lessening in strength as nature accustoms itself to the new feature in the landscape; removing the copper for a week or so every two months or adding a few hundred pounds after a few months to bring precipitation yields back to the desired level might be needed. The sudden change, when a quantity of copper is first placed somewhere, is always more dramatic than something in permanent position. Winds will show a marked preference for the direction of a row of copper tubing,even the Jet Stream will align to it. An acquaintance once said he saw an article in a mechanical magazine where they discussed using rows of copper around airports to assure that planes that are landing or taking off are facing a headwind rather than no wind or a tailwind.

Tornadoes and hail are problems that could be improved upon. As has been noted, there is considerable variation in the kinds of arrangements whereby copper tubing can be advantageously positioned; eventually enough different kinds of experiments will have been tried with each evaluated as to the effectiveness of preventing tornadoes, and hail.

For example, the litchi tree problem of needing as little wind as possible is probably solved by a slightly off vertical placement of copper, with a slight angle from the base to the top toward the west of about 20 degrees. Thus, the wind obeys the prevailing westerlies and at the same time the countering force of the copper slanted westward works in the other direction, causing a pronounced lull in the wind, lasting, no doubt, until the copper is removed. One would need to try a few different angles to the position of the copper and observe the effects, with the best angle eventually revealing itself.

As for how advantageously the farmers who are trying to grow litchis have positioned the farms,probably most litchi farming is done in valley areas. The valley below the copper would see even less wind than the westward slanted copper in a higher location nearby. As for what arrangement would most likely bring an end to the "Tornado Alley" in the middle part of the United States each spring and summer, perhaps more rain in the desert southwest and large scale copper arrangements near the north pole each spring and summer would help.

The polar regions could serve as huge conduits for water vapor and static electricity on a year round basis. A permanent two ton row of copper on the Northern Canadian Steppe or Northern Alaska might just send enough weather phenomenon north toward the North Pole, leaving that much less in the middle of the U.S..

Polar conditions being as they are, the logistics of placing some copper tubing there in a high location may prove difficult, but not impossible. Once placed somewhere exposed to the elements, it could be buried in snow and ice in three and a half days! So, a crew would have to visit occasionally and clear away all the frozen water.

Usually what happens with tornadoes is huge storm cells develop out of thunderstorms, and these sudden buildups are hard to predict. A kind of copper arrangement could be tried where the angle was more to the horizontal to some adjustable extent with the wind, and cause winds to be just robust enough to keep a storm front moving without lumping into huge cells, but still yielding precipitation, and this might succeed in preventing tornadoes.

Perhaps a multiple copper placement in the entire area, four to eight different locations spread several hundred miles apart, that together with helping more wind prevent tornadoes, also assured that individual storms are decentralized enough? Unfortunately, experiments with small amounts of copper in multiple locations would be difficult for one individual to do. With eight different locations with around 400 pounds of copper spread from Wyoming to Alabama maybe one wouldn't get huge cumulo-nimbus clouds, but a more generalized cloudiness and rain without tornados and hail. When cumulo-nimbus start to look threatening, several rows of lead sheets might bring down the severity of upcoming storms considerably. The middle of the United States in spring needs rain, though. Quite a bit of agriculture goes on there. Trying to stop all precipitation altogether with lead wouldn't accomplish much either. The absence of tornadoes would be tempered by the absence of rain, and drought.

Recounting what I have observed will happen with an experiment involving placing copper coils in circular stacks in a high location starts with placing the objects under a cloudless sky. One hour later,some small white cumulo-nimbus clouds will begin to appear just to the east of and right above where the copper is resting. After six to eight hours puffy clouds will be appearing on all sides. At around twelve hours an additional cloud formation will begin to appear, surrounding the area where the copper is. This cloud is a generalized hazy cloudiness, that thins out the farther from the place where the copper is resting you go.

If the experiment started at 6AM, the first night wouldn't reveal much, but the morning would show increased cumulo-nimbus clouds on all sides continuing to grow larger and thicker, especially nearest the copper. Soon, any trace of where the cause of the cloud development was would be hidden by complete overcast, as the entire cascade of lower pressure continues. One must realize that if a path of least resistance exists where the copper is, it has a field, an area where it is most intense. Schooling particles there will be compressed closer together over time as the higher pressure outside the area of the copper's most intense output exerts a push.

Sometime in the afternoon of the second day the sky will have become completely overcast. By dawn of the third day,when the copper has been around 72 hours in its location, precipitation will develop.If it is during the warm months thunderstorms would be likely. If an appropriate quantity of copper is used, 10 to 12 hours of rain off and on will come and go, followed by a cold front, with winds blowing down from the north behind the low pressure system. Too much copper and the rainfall is more intense, and lasts much longer.

Copper on a mountain and clouds above?

Leaving the copper in place, assuming that the location is in the northern hemisphere, after a day or so of northeasterly winds and colder temperatures, the wind swings back to originate in the southwest, and cloud development begins anew. On the first day after the cold front, the probable location of the copper would be apparent again from the area of generalized cloudiness, where the cumulo-nimbus clouds are the most intense. The second day again will see complete overcast take over, and rain could begin a little sooner than the third day after the cold front, seven days into the copper being where it is.

This pattern would likely continue indefinitely, with

occasional two day periods of steady rain with breaks between downpours once in a while. If this happened to be Lake Chad, and the residents were intent upon refilling the lake, the choice to leave the copper more than a week or two might be made. In most instances the copper would need to be retrieved, brought closer to sea level and sequestered in a building.

Explaining all that might happen with copper and the atmosphere would take a lot more experimenting and observation than I have accumulated. There could be hundreds of different arrangements one could try, each yielding a slightly different result. One would have the complexity of the topography of the Earth at any one location, along with normal weather patterns in the area to consider.An experiment in one location on Earth repeated identically in another location could yield surprisingly different results. The only way to find out how much copper to position somewhere would be to conduct experiments and build a database containing as much detail about previous experiments as possible. Information such as distance from the nearest ocean, altitude of the experiment, quantity of copper and what type of arrangement it is in, all these things would serve as a guide to future experiments. That way, disasters could be minimized. Currently it appears that flooding could indeed become more of a problem, with the authorities unwilling to concede that effects are produced in the atmosphere by copper in a high location, and the first edition of this book being available to the public for over a year.

One thing that needs mentioning, when weather modification activities are planned, is the consideration of an escape plan. This applies to any attempt to modify the weather with copper. For instance, the county of residence where you live decrees a weather emergency, rain is needed, and you and two other people are entrusted with placing some copper tubing in a high location previously chosen. You and your two colleagues proceed to drive a truck loaded with copper to the agreed upon site. For example here in the Las Vegas area, that could be Wheeler Pass, snaking up the west side of Mount Charleston. Once the truck leaves the Blue Diamond Highway, it is dirt road for twelve miles up, with some hairpin turns and deep ruts.

You arrive at the chosen destination and deploy the copper as described by the meteorologist; three days later its starts raining heavily. You receive a phone call after 6 hours of hard rain to pack up the copper and return to town. The dirt road is a quagmire, all the dried up river beds are now active rivers. There is no hope of driving the truck down the dirt road with the rain as it is. With the situation just described, the copper would almost have to be removed by helicopter, and that could be dangerous as well if there are electrical storms.

Thinking that one is in control of the weather, getting more rain than bargained for can happen,and without a way to get the copper back down the hill, can lead to flooding, and ruin. Dirt roads and arroyos can change dramatically with precipitation. Finding a location that is safer and from whence one might leave safely should be a priority. Maybe that location won't be as elevated a position as the original, but would still work, maybe with slightly more copper than the higher location on the dirt road to achieve a similar result. If all one accomplishes is dangerous weather, damage to buildings, farms, what would be the point? Control is different from runaway disaster. When one is in control, disaster doesn't happen. It would be important that the exposure of the copper could be ended on short notice, and returned safely to a building at a lower elevation.

The start of the experiment begins with those entities in the atmosphere that are responsive to the electrical field of the quantity of copper in a high location beginning to proceed in that direction. The water molecules in the atmosphere, and the hypothesized primordial specks begin to align themselves along the path of least resistance. Relative humidity begins to rise. Accordingly, the concentrations of primordial specks also begins to rise. As more water molecules align themselves along the path of least resistance they exert increasing gravitational pull on the electrically inert atmospheric components.

Relative humidity is derived from the humidity at the saturation point, which is 100% relative humidity, and the actual quantity at 30 degrees Celsius is 30 grams per cubic meter, a little over an ounce per kilogram of atmosphere. At 20 degrees Celsius the saturation point is around 15 grams per 1000 grams of atmosphere. The actual amount of water in the air is only 1.5% by weight at 68 degrees Fahrenheit, and yet, at that point, rain would develop. At this temperature, then, 7.5 grams of water per 1000 grams is 50% relative humidity, yet only 0.75 percent of the air by total weight is water molecules.

At the outset the estimated distribution of primordial

specks in the atmosphere is around ten times the mass and energy of the known atmosphere. If these mysterious entities were to increase in concentrations as well as the water molecule, the nitrogen and oxygen surely come along as these quantities of things all traveling in the same direction increase over the time of the experiment. What concentrations primordial specks reach could have no physical boundary beyond which it could accumulate further, as water molecules do. At 100% relative humidity, water molecules condense out in the form of precipitation, but primordial specks could still be increasing. Or it could very well be that a threshold exists where concentrations are sufficient for primordial specks to begin condensing into hydrogen, and that threshold is reached some time during the 72 hour experiment. Both entities responsive to the electromagnetic force of the copper may begin condensing out. However the two combine, together they are enough to pull the electrically inert nitrogen and oxygen along the path of least resistance. Thus, barometric pressure falls for the reasons mentioned earlier, everything airborne traveling in the same general direction. Water molecules are lighter than either paired isotope of nitrogen and oxygen, so 3% water by weight might be a higher percentage by volume, perhaps as much as 5 or 6%. Water by itself would naturally rise and form clouds after evaporation, being lighter than the nitrogen and oxygen isotopes.

Chapter 5. Public Concerns

"Too Many Cooks Spoil The Broth"
Old Adage

There is a definite relationship between how much copper is placed near a mountaintop and how much precipitation occurs. One could deploy more copper than is necessary and cause flooding or tornadoes, possibly even hurricanes. Therefore,this discovery could be used as a weapon by anyone with less than noble intentions. This could very well be the reason why no meteorological text contains any information about this discovery. It is simply too easy to cause changes in the weather, changes that could prove destructive in nature.

A ton of copper could leave three states underwater in about four days. Never should anyone attempt an experiment such as we've described with that much copper, unless they are in the Gobi desert and have already tried a half ton with little success, or are trying to extinguish a forest fire. This is the moral high ground that the scientific community, the media, and the government are clinging to desperately, about to be swept away by the tides of progress.

While in 1910 or thereabouts it may have been well and good to sweep this discovery under the rug, so to speak, and pretend that it doesn't exist, in this current era of rapid technological advances, world-wide communications, and the current problem of global warming, this simply won't do. What do you tell your ten year old child when he/she discovers this on his/her own? What is the attitude of a group of teens toward authority figures after they have found this out on the world wide web? How does the current generation of adults appear to them, other than hopelessly inept? It is apparent that the vast majority of meteorologists currently practicing that profession have probably never even heard of this possibility, except perhaps for one day, half asleep in a classroom, it was mentioned briefly and debunked erroneously. Some members of the meteorological community must have in the past been aware of this, and chose not to communicate what they had discovered.

The same holds true for people other than meteorologists. A few engineers surely learned this at some time. The early twentieth century was the age of the engineer, and it included three decades of wet weather,ending abruptly the spring following the Stock Market crash. Besides copper having an impact upon the flow of air molecules, a non-conducting element such as lead would have the opposite effect, logically. Peace loving humans are already in possession of a means to counter attempts to disrupt peaceful weather conditions. At the first sign of possibly severe weather lead could be placed in a high location so as to bring about higher air pressure, and consequently, less severe weather. In the United States there are countless barometric stations maintained all over the continent.

At the first sign of an unusual drop in the barometric pressure in a given area, one that is unplanned, a helicopter search for the offending material could be conducted, and using a barometer, readings could narrow the search area down to within a square mile or so. Since the material in search of would have to be in a high place to have any impact to begin with, I feel confident that finding and removing unauthorized attempts at modifying the weather could be accomplished within a few hours, far before the seventy two hours needed for a storm front to fully develop.

Where did I leave that copper?

In the event that someone were to burn copper compounds after dark, Charles Hatfield style, lead might be the only recourse, though pinpointing the probable source of these copper compound plumes could also probably be accomplished, as for instance, when a substantial drop in barometric pressure sets in motion a search for a quantity of copper and nothing is found.

Wherever the barometric pressure readings were lowest would probably be fairly close to where the copper compounds went airborne. The area could be put under surveillance until the next time the culprits attempt such a misdeed. Ideally, only licensed meteorologists with the appropriate qualifications and permits would be legally conducting interventions into the weather in an area. Any other attempts by persons unknown could and should be considered an act of terrorism punishable by death. In contrast, consider the situation as it is now. Anyone with a pickup truck and access to enough copper or lead, or the funds to pay for some, could make life miserable for a lot of people.

Soon, I hope, barometric pressure will begin to be watched more closely for unexpected changes. Places that are prone to disastrous weather should be ready to position lead on high from time to time. This is not something I ever did. Nevertheless, the laws of physics are on my side, and extrapolating what is likely to happen when lead is placed somewhere high up from what happens when copper is placed somewhere high up shouldn't be impossible. Could be placing lead high up has less of an impact than using copper, and a war between the two metals would see copper emerging as the victor.

The only way to find that out would be to begin some experiments in this regard. If it turns out that flooding can't be stopped by inserting lead somewhere high up, then one would be forced to find the offending copper and remove it. By remove it one would take it to a location much closer to sea level and house it inside a building. There is also the possibility that negative electromagnets would repel water molecules, making cloud development difficult, if not impossible.Positively charged electromagnets would likely attract water molecules, and experiments with those should also be carefully monitored. The heavens can open up with scads of precipitation, and it can stop raining for months, both through human intervention. The human intervention wouldn't be evident, though, without a concerted search, and the difference between a naturally occurring event and one with an added metal deposit or electromagnet wouldn't be evident either.

A swift response coupled with a harsh penalty for such acts could make things far safer than they are now. Hurricane Rita and the aftermath featured a number of web sites that claimed this storm was a man made event, though there wasn't much elaboration as to how this was accomplished or by whom. There was a reference to electromagnetic pulse generation, such as the experiments conducted with H.A.A.R.P. Here we are, then, totally unprepared for such acts in an era with such a thing as the world wide web.

A quantity of copper on a mountain top could be pinned down fairly accurately just by watching as the clouds develop. The area nearest the copper will have, in addition to some thicker clouds than the surrounding area, a misty, hazy cloud structure enveloping a one to five mile circle. Farther out, the clouds will turn into thin streams, all seeming to drift toward the area of thick clouds with the misty, hazy envelope.

To watch clouds one day, and see clumping clouds of the cumulo-nimbus type to the south,and clear skies to the north, with a flow of clouds welling up from south to north, if you watch the northernmost clouds you will see them slowly disappear before they fill the whole sky. That is a pretty good indication that lower barometric pressure is taking place to the south, and if one journeyed south to where one finds the thickest accumulation with accompanying misty, hazy cloud structure,that would be the best place to start looking. Rows of cumulo-nimbus clouds are also a fair indication that somewhere along the row is its cause.

The civilized thing to do would be to have one particular agency in charge of the weather in an area, and prevent any other interference lest weather events go awry. Experimenting with certain amounts of copper, or lead in some particular area while other experiments of the same type are also ongoing, while not knowing what type of experiments these other experiments are, isn't going to help establish accurate guidelines to obtaining precipitation in the desired amounts. A country should have the task assigned to the meteorologists of the country, and for the reasons pointed out, the meteorologists should be the only agency involved.

Since no experiments can be devised on a small scale to replicate what might happen in the real world, the real world is the only place experiments could logically be conducted. Unfortunately, in the real world there are people living in the area, buildings, livestock, planted crops, and wildlife. Scientists would be forced to exercise some restraint lest storms prove too severe. One couldn't place an excessive quantity of copper on high and stand there with a notepad recording the results while the entire area floods out.

To begin, all the meteorologists of whatever country need to do is determine where precipitation in the country could increase some, and where it could decrease some. If no water shortages or flooding are happening, and that could be possible in some countries, the weather modifying agency need do nothing. When conditions change, and some provinces begin to run dry, the experiments cautiously begin.

This difficulty, that any experiments of the type we have been discussing would have to be conducted live, in the real world is the second major impediment to scientific investigation. I watched a University of Southern California Meteorology discussion on their television channel, and the speaker kept stressing to the audience that all the ideas that people were coming up with about meteorology shouldn't be acted on in the real world. Don't do anything, he said several times. It appears it is hard to coax them into the real world when there are computer simulations to try, and conferences to attend. Academia is where they plan to stay, and the dangers of this precipitation making process described herein and the necessity of experiments running live in real time make this whole notion a very hot potato for the meteorologists. No one is eager to take the first step.

Lead atoms are quite unlike copper atoms. The arrangement of lead atoms features a very haphazard distribution, with no easy path for electrical current to follow. The friction encountered by electrical current makes it impossible for current to pass through lead. Tiny gas molecules and other atmospheric entities find no easy path to travel along in the vicinity of lead, so the barometric pressure in the vicinity of a quantity of lead would tend to be higher since the gas molecules are encountering each other more frequently, having no smooth path to follow.

Naturally, too much lead put in place in a high location would lead to drought. This too, should be considered an act of terrorism and be strongly discouraged by the authorities. Barometric pressure readings should also give investigators a starting point in the search for outlaw lead placement as well, the difference being that the search would be centered on those areas with higher barometric pressure rather than lower. We can only hope that this whole idea doesn't escalate into a war between competing quantities of metals.

A large negatively charged electromagnet might also make cloud development difficult. Finding an electromagnet responsible for undesirable weather might prove a little more difficult than finding a considerable accumulation of copper or lead somewhere high up. Nevertheless, when severe weather begins to erupt, or a drought begins, these possibilities should be investigated. Perhaps a strong electrical current is detectable from a certain range with some sensors.

Right now we are helpless against any acts such as mentioned, as we have shown. Until the entire human race is able to learn of this discovery in an encyclopedia, the playing field will not be level, so to speak, and the risks are greatly increased with the situation as it is now. Human beings can adapt to changing conditions and new discoveries. I doubt that there would be many problems once the initial first steps are taken to begin a system of monitoring the weather that would include active participation in the making of, or breaking up of, storm fronts.

If you look up weather modification on the internet you will find that several companies already exist claiming to be able to assist with drought mitigation. The alleged techniques that these companies use varies. A lot of it seemed pretty unscientific. There is a lot of mention of electromagnetic pulse generation on the internet these days.Some way of quickly and accurately finding such EMP broadcasts should they pose a threat to peaceful weather conditions should be explored. Of all the possibilities, EMP type broadcasts may prove the hardest to track down, but since it is also the most expensive and difficult to construct, other methods would find more use.

The United States and Russia each have a place where these EMP pulses are generated. The one in the United States consists of hundreds of 700 foot tall towers laid out in a field several football fields in size. It would be quite difficult to hide something like that if some private citizen were to try to build one. The day has not arrived when something of the strength of the electromagnetic pulses emitted by HAARP could exist in a hand held device. Those that exist in the possession of governments now can fire at will just about any-where and not be held accountable for weather misadventures that result. There may be other players besides just the United States and Russia.

Ideally, some further investigation into this phenomenon will be conducted, hopefully by meteorologists. The possibility that any experiments conducted could be sabotaged by other persons with some other placement of metal in some nearby location needs to be considered. If secrecy as to the time and location of a proposed experiment must be maintained in order to assure the purity of the experiment, then the public won't know when such an experiment is conducted. Therefore, the public will have to trust the meteorologists involved in so far as faithfully recording what meteorological events take place.

For example, the meteorologists plan an experiment with 500 pounds of copper. An opponent of this idea becoming public knowledge learns of this proposed experiment, and places a ton of copper on an adjacent mountaintop, or building roof, as the case may be. Three days go by, and the resulting storms are now too severe. The meteorologists conducting the experiment decide not to release the results of the experiment they conducted.

Or, better still, the meteorologists plan an experiment with 500 pounds of copper. An opponent of this idea becoming public knowledge learns of this proposed experiment, and places a ton of lead sheets across a nearby high location. Three days go by, and little or no precipitation occurs. The meteorologists then report that the experiment failed.

It would be imperative that these experiments be conducted with as much secrecy as needed to reduce the likelihood of sabotage. Only then would the true cause and effect relationship reveal itself to observers. Currently one would have to view HAARP as a possible saboteur of weather experiments here in the United

States. Whether those conducting the experiment are answerable to anyone and whether the results will be reported honestly and accurately during these experiments one can only hope. Naturally results of these experiments wouldn't be learned by the public until some time after the experiment happened, which would be fine since the experiment was secretly conducted.

In the final analysis it may prove more difficult than one would expect to validate what has been asserted here by properly professional people in a properly professional setting. Obviously, if anyone over the years had happened to notice that this process in nature did what it did, they did not communicate this to the general public, or were prevented from doing so by a paranoid and overly militaristic government. Currently, that same government could make investigation of this phenomenon difficult if HAARP is put to use thwarting weather modification experiments.

I don't know why this process in nature isn't already in the encyclopedia. Many objections were raised by persons with whom I've spoken to the effect that if such a thing were true we would already know about it, therefore it can't be true. That is hardly a satisfactory conclusion either. The possibility that the truth is still to be discovered with regard to this phenomenon still holds true, even if it would seem to most that such a thing would already have been discovered. To assume that HAARP will be effective one day as a device to modify the weather with, one would also have to assume that the electromagnetic waves that propagate from purified metals would also have some effect.

The machinery creating electromagnetic pulses would also broadcast waves in the electromagnetic spectrum. So depending on the wavelength, high or low pressure could be created by these pulses. The HAARP experiments involve firing concentrated pulses at parts of the ionosphere, not just sending out random pulses at some wavelength, though it could probably do that as well.

Concerning the impact of this discovery upon world markets, it probably wouldn't be severe enough to disrupt agricultural futures. After all, the mere fact that humanity has discovered one more thing to help them survive happens all the time. Agriculture won't suddenly become the easiest thing to do on Earth. There will still be work involved in growing things, caring for them as they grow, and bringing them to market after harvesting. Pests that threaten certain crops could proliferate in a wetter environment, making the growing of some plants even more difficult than previously. And, as mentioned earlier, not everyone will learn of this process, and those that do wouldn't be a significant portion of the population for some time, even if the media condescended to say something about the discovery. Years would have to elapse with the process included in encyclopedia before a sizable portion of the population is aware of it.

Things might just go on much the same as before, with some significant differences. No one would be wasting time trying to discover something that has already been discovered, famine would cease, houses would remain intact longer, forest fires would get put out safely and efficiently, fewer ships would be lost at sea, people would be happier, etc. The overall economic impact would probably be quite good, and it wouldn't happen overnight. Some of the economic ramifications of minimizing weather disasters and having a world economy using abundant water throughout would take years to be noticed.

Businesses might exist after years of peaceful weather that would have stood no chance of getting started without the stability of the weather. Had mankind not taken charge of the situation, the resources that enabled these marginal ventures to succeed would have been entirely taken by the construction industry and the rebuilding of structures destroyed by weather and the building of desalinization plants, and they simply would not have existed at all.

New or expanding businesses could start out with the realization water could be made be available in the area in which they plan to do business. That already changes the business environment of many countries that have in the past been prone to water shortages. Businesses with plans in underdeveloped countries might have to do their own precipitation making, or explain to those responsible in that country how it could be done. It would also change a lot of the prospects for desert and semi arid regions in more well developed countries. Arizona already has some extensive irrigation systems built and ready. Add water.

The likelihood of anyone profiting monetarily from this idea is little, other than a rise in the standard of living that would be slow at first but gradually increase, as one thing leads to another. Certainly there are instances where abundant precipitation could produce profits, but it would be the things grown and sold, not the precipitation itself, that produced profit. Though there are some expenses involved, it is the most economical solution to a plethora of problems concerning the environment around the world. The arrangement of six foot tall copper coils tied together with straight bars of copper tubing described earlier isn't necessary if one has a flatbed truck with a lift feature similar to what a garbage truck uses. Lay the tubing along the bed of the truck in a row tied down securely with copper wire, and park it in a favorable spot. Elevate the bed of the truck to as nearly vertical as possible, facing west. Wait for three days.

Naturally anyone who has an idea they are promoting thinks it is the most worthy of public funds or should have persons interested enough to invest money in it. In this instance I am no exception. Nevertheless, the problem with this discovery is that no one has any facts or experimental results that one can point to definitively and say, there, publication such and such, pages so and so,etc., where some specific persons conducted specific tests with exactly so many pounds of copper and so on. The truth is really still to be determined, at least as far as I can see. That meteorologists scoff at the idea as being long ago disproved doesn't convince me that this isn't some kind of trickery, involving superior minds thinking that the lowly masses shouldn't know about these things.

There is no doubt in my mind that my hypothesis will be ultimately proven correct. I think that if a meteorologist read my first edition, he would concede that things probably do happen when copper is placed on high, but probably disagrees as to whether it will always do basically the same thing. Hiding behind chaos theory doesn't change the imminent usefulness of strategic copper placements or its beautiful simplicity, and as scientists, with the final say in matters scientific, doing that fails the human race, and impedes progress.

If, once the truth is determined, no one sees fit to spend time or money on an idea whose fruition will not be monetary, but rather enhancements to all life on Earth in the long term that each individual will receive in small increments that will largely go unnoticed, such as not worrying about the weather each and every time one steps outside, let the market do what it will. I cannot force my value judgments on anyone nor do I intend to. I just hope the science of meteorology means the same thing to me as it does to the other 6 billion plus humans present on Earth.

Besides, large scale changes wouldn't take place overnight. No one is expecting immediate changes. What is hoped here is that enough people share my curiosity and desire to see this process described herein get more attention by the scientific community. Some time will surely elapse between the publishing of this book and the next logical step. That next logical step would be seeing a science presentation on public television about some of the possibilities raised in this book. The only way that can happen is if some scientists actually do investigate. Once that happens the human race will have cleared a major stumbling block and be on the way to a much brighter future.

Human beings are under the necessity of sharing the Earth with fellow humans as well as all other living things present on the planet. That is generally speaking. We all know there are certain creatures most humans regard as vermin and would prefer not to share anything with. We all know that science deals with ascertaining what things are, and how things work, so if and when we find something that works that may prove harmful we record the fact and the governing body steps in to regulate the use of such a discovery.

The Bureau of Alcohol, Tobacco, and Firearms exists to protect the public from the dangers of these things. There aren't just a few people with guns in hiding and everyone else pretending guns don't exist, right? Guns are a device we are forced to live with everyday because someone discovered how to make one at one time and that knowledge, with improvements, is now part of our civilization.

If the ATF became WAFT, weather, alcohol, firearms and tobacco, that would simply be the way things are. Whether or not it is economically feasible to modify the weather,if we once and for all declare publicly that we are capable of it and it poses a possible danger, one may rest assured that the government will step in and enact legislation to protect the citizens from possible harm. Marconi stole Tesla's ideas regarding wireless transmission of radio signals, so ruled the Supreme Court, before I was born. It happened that Marconi was one of Tesla's lab assistants during one of Tesla's more prosperous times and that is where Marconi got the ideas he claimed were his own. I rather resent having learned that Marconi invented the wireless in school, almost two decades after the Supreme Court decision. This is just one more instance demonstrating how sluggish governments are when confronted with change. I sincerely hope there are new textbooks now with Tesla credited with the discovery of wireless communication. Could be that some bureaucratic decision was made to exclude any mention of Tesla in public schools. I never heard of Tesla until some time after graduation from high school.

What's the verdict? Do we draft WAFT or are we daft?

Most of the countries known as Third World countries already have the problem of either not enough precipitation as a general rule, or too much precipitation, or both, at various times of the year. The weather problems they experience is a large part of the reason why these countries are classified Third World countries in the first place. Bangladesh, for example, would benefit from the even playing field once this knowledge is assimilated and, ultimately, put into Encyclopedia for the curious to learn. Simply the absence of tropical cyclones hitting that low lying country would in and of itself be a great boon to them, as well as their being able to lessen the impact of the yearly dry seasons, which could be easily accomplished without umpteen meteorologists and weather monitoring stations.

It is possible that some countries who were at odds with each other in Africa had positioned lead in large quantities so as to cause neighboring countries to suffer from prolonged drought. One other possibility that needs pointing out, lead bullets scattered all over areas where skirmishes took place between rebels and government defenders will behave like real lead, and believe it or not, could be responsible for drought. Africa in 2008 was in the grips of some of the worst flooding it has seen in a long time. Now, maybe the metal of choice has changed between the warring nations. All we can truly say is that the flooding in Africa in early 2008 was one more instance of weather changing to wet abruptly.

That is the entire point of this presentation. Time and time again some anomalous piece of weather information stares us in the face and adds to an ever growing total attributable to which cause, global warming or man made metal deposits? A lot of these instances will never reveal their true cause. A fair proportion of them are probably man made, though. Wisconsin tornados in January 2008,for instance,almost certainly got a little help from some man made device. The same three counties in Wisconsin saw tornadoes again in October of 2010, and again in November of the same year. Southeast Wisconsin is not known for tornadoes even in spring or summer. I lived there 39 years and never saw one. Did someone I spoke to back then try an experiment using too much copper? No one is investigating what caused those storms. If human hands and a quantity of copper were involved, no crime was officially committed, except perhaps the failure to request or report a weather modification experiment to the authorities in Wisconsin, if they have laws requiring permission or reporting.

There is a problem though, isn't there? The official stance concerning weather modification involving quantities of copper in an elevated location is that it doesn't do anything. Therefore, if someone had placed copper somewhere on high, that had nothing to do with subsequent weather events, and was just a coincidence (again). This one is a pretty strange coincidence, since the area, near Lake Michigan, almost never sees tornadoes in spring and summer, and now tornadoes have occurred there during the cold months on three separate occasions. Nature is certainly capable of any and all weather events or they would not have happ-ened in the first place; what is unresolved is whether they would have taken place without some additional man made factor entering into the equation.

In light of all that has been discussed, watching any program about nature or the environment on television is quite a distracting thing to experience considering that so many suppositions that are made on the show are negated by the possibility that we are fully capable of doing something about it. No full discussion has ever been made of these topics with the assumption before hand that weather modification is possible. Even a simple nature program can offend the enlightened, as for example, a film about certain areas of Africa and the wet and dry seasons and the animal migrations. "And then the rains came." runs the narrative, and at that point one finds oneself wondering what the intelligent beings who lived there were doing prior to the rain arriving.

A lot of the global warming documentaries are similarly flawed. A major theme running through most of these are that global warming will change weather patterns drastically, such that wet areas become dry and vice versa, storms will be more severe, etc. That remains to be seen, especially when a means of changing things by human hands exists and has for well over a hundred years. Another recurring theme in discussions of global warming is that we will reach a point of no return one day where runaway global warming will takeover and be unstoppable.

I hope someone takes notes on how many weather events destructive in nature occur after this discovery is placed in the encyclopedia,and compare that total with the totals that occurred before that for a comparable period of time. When a 98% reduction in destructive weather events happens, how wise will keeping a process such as this secret seem? We are at the top of the food chain on Earth. We are it's custodians. We must do the weather, it's as simple as that.

The only way to resolve such problems would be to establish guidelines as to how much and how often precipitation can occur for reasonable amounts of water to be available. It is not too difficult to perceive when an area is experiencing drought.Plants start to shrivel in the dry heat, rivers run dry, animals begin to die, and the necessity of precipitation is evident.

Planning ahead and assuring that such conditions do not occur would be a reasonable course of action. When rainfall becomes excessive and flooding starts to happen is also not difficult to establish, so finding the happy middle ground where neither drought nor flooding occur shouldn't be too difficult, nor should actions taken to achieve such goals be subject to litigation by unhappy persons discontent with rainfall totals, if as we expect the authorities assign these tasks to meteorologists, provided no damaging weather occurs.

How many instances of damaging weather will result from initial experiments before the experimenters get it right? This device and how it can be used, and to what quantities safely, and what quantities pose a problem has "proceed with caution" written all over it. Any competent and thorough experimenter would begin with a safe quantity, perhaps three or four hundred pounds, see the cloud development and maybe a little light rain,pack up, return some time later with more copper, find the storm system to be of the size he is aiming for, and assign the task to a local crew on standby for the weather service.

Eventually the entire country will have standby crews with well established plans in the event of drought, or flooding. Before that can happen, though, this theory would need some discussion, and it remains to be seen if humanity can overcome the differences of opinion they all have regarding this issue. One thing is certain, though; ignorance has not been bliss.

There are other implications of this discovery that go beyond the usefulness of the discovery itself.

Philosophically, this discovery would not further the goals of those who believe that a God created the universe. Once again, it would point to the sad fact that we as humans are quite on our own when it comes to dealing with things, including the weather, the various manifestations of which are already classified as "Acts of God" by the authorities.

As the Boy Scout hopefuls pointed out so aptly who were excluded from joining the Boy Scouts because they were atheists and then sued, we do not have His signature on anything. It has been noted that there has never been a President who did not side with the notion of a creator. So the mainstream thinking is not going to find this discovery useful in forwarding their agenda.

My problem with that is truth being lost in the shuffle. In the true sense of the word, the notion of a creator is still an unproven assertion. Besides that, the whole notion of faith is flawed. The famous quote goes, "To Doubt Is To Think". We as a species are not faithful beings, at least not in the sense of ascertaining the truth or falsehood of something. Certainly a husband can be faithful to his wife. Human beings are curious by nature, and learning what is true or false is important. We are the only creatures known that are capable of thinking. The very definition of man is "thinking animal". To have faith in something that one has no proof of is contrary to our very nature, and a direct insult to a God, if there were such a being.

Leaving no stone unturned in the search for God turns up the most enormous universe perhaps, an unknown number of times bigger than our known universe, looking and acting like a wilderness in every respect, if we assume that what is beyond here is more of the same as we have here, with some really big black holes scattered around that we probably don't have here because two collided in this tiny quadrant of the entire universe to create our known universe and the nearest really big one is probably the "Deep Drift" object.

No sign of God in a known universe that contains an estimated 100 billion galaxies. Our Milky Way has somewhere around 100 billion stars. Considering that it would take quite a bit to create 100 billion galaxies and still more to create unknown multiples of that, God would certainly grasp our rules of evidence and respect them, these rules of evidence having been invented by our minds, which, if he exists, he created. Knowing that we are curious, able to doubt, and have limitations, he would have certainly signed a document for us, maybe even have explained in the King's English where he has been.

Probability of actually proving God's existence much lower than winning Powerball. Thus, there is a very slight chance the faithful are offending the supreme being, and an overwhelming likelihood they are wasting their time. I'm only presenting this evidence for the sake of completeness. Certainly anyone who decides to believe in whatever God is free to do so. That which never bore fruit can be abandoned, as well, though, provided there is freedom of choice and access to pertinent information. Obviously, religion isn't going to disappear overnight because of whatever arguments raised here; it could become extremely unpopular in the not too distant future, though. Terrorists are already demonstrating what fairy tales do to people.

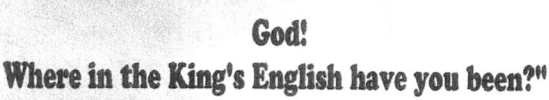

There is one aspect of religion, how it began, and when it began that everyone I am sure is aware of, and another that seems to have been missed. We all know religion began long ago, when only a handful of people were literate; the printing press came along in 1450 AD. Prior to the printing press books were reproduced by hand, one at a time, and in 3300 BC there were surely fewer literate people then than in 1420 AD. What survives must be tempered by the new knowledge of the present day.

The other thing, which most people probably wouldn't be aware of, is a rare cosmological event. A supernova of a common type, after the initial explosion,some time later will have a debris field surrounding a white dwarf that has very powerful rays emanating from opposite ends. These powerful rays increase in energy as they pass through the shock waves of the debris field. Though estimated to be a very rare event, the Earth could cross paths with such a cosmic beam. It is even likely the Earth could be completely destroyed from a direct hit by a cosmic beam if it came from near enough away. That leaves the question what would happen if one hit us from far enough away that it wasn't fatal.

Surviving from the old testament is the parting of the Red Sea by the hand of God, and the Noah's Ark fable, where it rained for forty days and forty nights. A supernova remnant cosmic beam from far enough away would have spread out over the light years it had traveled until when it came into contact with Earth, it could have been a mile or two wide. It would have been a bright white light, very hot, covering several hundred yards to two miles or more, coursing across the surface for five minutes or so, about the same time as an eclipse. It would probably melt a few rocks near the surface, besides incinerating plants and animals as it moved over land, while over water creating vast amounts of steam, some of it the Red Sea.

Those two events could have happened, the one followed closely by the other. Since all the steam that was created rose into the atmosphere, it would all eventually come back down, and take some weeks to do it. The forest fires created by the cosmic beam on land get put out by the rain that followed, and all evidence of the path this cosmic beam took would have been gone in a decade, except the aforementioned melted rocks near the surface. There are certain isotopes of certain elements that would be in greater abundance in a rock that had been melted by a cosmic beam such as this.

If some of those isotopes are found a few hundred yards from the oldest great pyramid amid once melted rock, and still more right near the shore of the Red Sea, that would confirm that the ancient story was based on a true event. The oldest pyramid is from around 3300B.C., around 5300 years ago.Going one step further after finding such isotopes the scientists would carbon date the once melted rocks with the unique isotopes, and my guess is they will prove to have been melted right around the same time, maybe a few decades before. Why write an account of the parting of the Red Sea, and of rain for forty days and nights, unless it really happened? What other earth shattering event would have inspired the building of the oldest and largest pyramid in Egypt?

Assuming such an event really occurred leads one to the realization that any witnesses to the event at the time would have averted their eyes since there was little doubt to them but that the bright light was the hand of God. Witnesses bowed down to the ground and covered their heads. Then, 4750 more years had to elapse before the development of the printing press.

Each reproduction of a book prior to then done by hand, one at a time. These reproductions were accomplished almost entirely by the priesthood. Rabbis, vicars, monks, and priests were almost the only literate people in the world back in 3300 B.C.,and that was still true in 1450 A.D. Some people other than the clergy were literate, surely, and that number slowly grew over the years, but the vast majority of humanity was still illiterate then. There are still countries where women are not allowed to receive an education.

Each of those reproductions of books about the religious events gets a different interpretation and possibly a slight change in the manuscript as it is reproduced by the clergy person, and as the years pile up and language undergoes a number of changes, the slight changes and the poetic license taken by the person doing the reproducing begin to add up.

As terrifying a sight as such a phenomenon may have been, it appears that the only thing that could have possibly done this, part the Red Sea, was an event triggered by some distant explosion of a star, and the resulting events following. This was not apparent to the people that built the first great pyramid near or on the site of this amazing event, the apparent hand of God reaching down onto Earth. It is unfortunate that as a species we had not developed the astronomical knowledge to correctly identify what this event was when it happened.

Were such an event to happen now, it would be correctly identified, probably predicted a decade in advance of the event, yet still cause massive loss of life and flooding. There was some investigation into the parting of the Red Sea that concluded it may have had the flow of water stopped temporarily because of an earthquake. Whether that conclusion was definite or just possible is unclear. The account was of a hand reaching down to Earth; for that to have really been true, an earthquake wouldn't explain it. One can be certain it wasn't a hand like a human hand, it was the hand of God. What does that signify,then, but a cosmic beam.

If some once melted rocks are ever found near the Red Sea to indicate the trail that a cosmic beam took that struck Earth, and there is some evidence that something terrifying happened back then,it better explains the religious fanaticism of early man. It is a stretch of the imagination to believe that the primitive peoples living way back then should have possessed some insight into the nature of things superior to the current accumulated knowledge of all the natural sciences.

What those people experienced may have been an exceedingly rare event, and the people then made assumptions about the existence of things just as we do now,in a world with much less accumulated knowledge. They decided a supreme being was responsible for that event. In other words, they didn't know what happened, and resorted to mythology to explain it.

An economics book titled "Human Action" by Ludwig Von Mises[8] has a part where the author discusses the notion of a supreme being, and the author points out that only discontented beings act.How could a being with infinite power want for anything or ever be discontent about something,and if that being ever were discontent over something that uneasiness felt could be forever removed in one fell swoop.Without limitations life no longer has meaning; a perfect being would never act, or at best have acted once. OK, God created the universe, but he will never act again, it would not be possible for a supreme being to be discontent more than once, therefore it is up to humans to think things through as though there were no God. Whether created or eternal, either way the universe is now a wilderness.

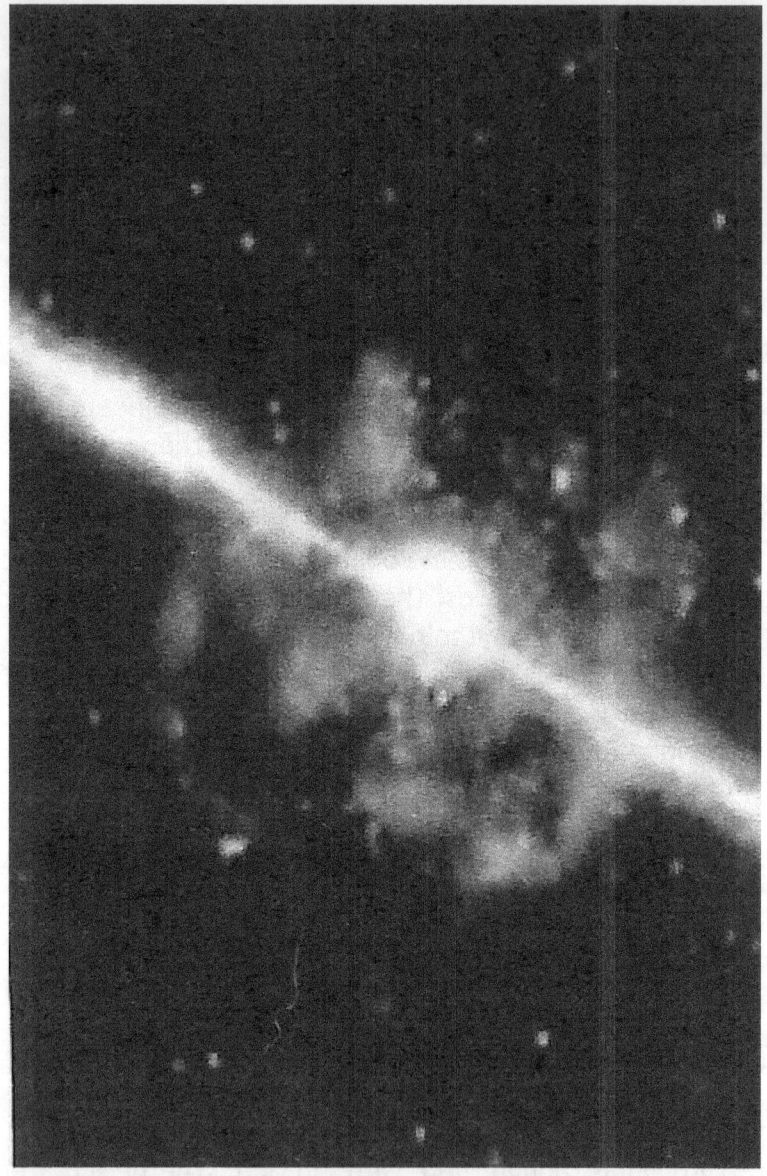

A one in 5 billion years event, we may have crossed paths with this

Lest the reader think I am getting too far afield with

the argument against religion, it was a subject that was originally planned to be avoided altogether. However, this book has more information that needs public awareness, and if somehow these issues never reach the public, one wonders why that is the case. Human society so often has just followed in the footsteps of previous societal arrangements and seldom exceeds those boundaries. The status quo is to prevent the public from learning anything new, lest it upset the old apple cart. There is little else to say other than that one learns something new every day.

The only other consideration with respect to arguments against religion that came to mind, and persuaded me to include these arguments, was the possibility that some of the youth of the world might eventually read this book. Anyone who has sunk his teeth into religion so tightly as to never let go will never change. But those who haven't already been indoctrinated have a chance of becoming thinking human beings, and making an informed decision about the nature of the universe.

God isn't necessary for humans to treat each other with respect. The obvious benefits of peaceful cooperation and free trade between countries and individuals, the development of culture and art by our society, movies, libraries, sporting events, anyone can see that it is in their best interests to be a cooperative member of society. That a small minority persist in defying common sense and committing crimes is undeniable. That small minority should not be any detriment to ascertaining the truth of matters of physical laws, or whether such processes described in this book do what has been claimed.

It may very well be true that a whole new type of vandalism has been introduced to the typical miscreant. That is unfortunate. Flooding can be much worse than drought, and bringing an end to flooding might start becoming a drastically needed thing. Flooding can be stopped, but this idea would have to be understood for that to be even tried. If humans could learn this in the next two decades, by the end of that time flooding is under control and virtually non existent. Wouldn't it be better to face those few years of flooding caused by vandals now, instead of forty years further into the future? Once past the first few years of learning of this, each country will have weather plans of their own, and precautions in place to prevent modern day weather vandalism. Supposing what I have presented to be plausible, there is really no other way to change the status quo, and begin the process of learning all this but one day at a time, and vandals will do their work long before the authorities do theirs.

Lead in thin sheets from north to south in a high location seem the best bet for stopping storms and ending flooding. These types of metal placement may become necessary. The more people are aware of both possibilities involving placing metals somewhere high up, the sooner impending flooding can be nipped in the bud.

God isn't necessary to develop laws. Common sense will suffice,the "reasonable man" concept of law could serve as the basis for human activity without any deity involved. Right or wrong is also easier to get to without the interference of some hypothetical supreme being.

God will have to take this discovery on the chin, as it becomes obvious that the weather is subject to natural laws only, and that humans can gain control of them.

If enough persons favor it, religious and historical holidays could be abolished altogether, and instead, a three day weekend is assigned to the last weekend of every month. The Post Office and other government offices, banks, and many corporations honor the three day weekend. Gas stations and restaurants often stay open on holidays, nothing would change that substantially. Why disrupt the flow of commerce with a Tuesday holiday? Simplification in this regard has already begun to creep into current holidays, some of them are being moved to Monday. Thanksgiving is another instance of less than thoughtful planning, Thursday is not a good day for a holiday.

When I worked in wholesale food distribution, the following Friday after Thanksgiving was one of the most difficult days of the entire year. Restaurants that order supplies daily had to wait a day. Other restaurants that order on Tuesdays, Thursdays, and Saturdays now needed to be supplied Friday, following the holiday. They were added to the regular Monday, Wednesday, Friday routes. As a delivery driver supplying wholesale food to restaurants,I have worked several fourteen hour days following a Thanksgiving holiday. July 5th usually had the same effect. New Years and Christmas both falling on a Wednesday can really lead to a lot of work on the following Thursdays. Restaurants want fresh produce, and during times of holidays, even more of some items. With regularly scheduled routes that all produce suppliers have, occasional Mondays off instead of the haphazard holiday schedule that is currently in place would ease the work load at critical times.

Traffic congestion during holidays only adds to the extra time necessary to supply restaurants that could not be supplied on the day of the holiday, increasing the total time worked. Everybody doing the same thing during holiday periods intensifies supply problems to a considerable extent. Declaring generic holidays on Mondays and letting humanity do its thing instead of declaring a patriotic holiday midweek where people set off fireworks would work out much better. The fireworks are getting old, and every year, responsible for some fatalities.

The simple confirmation of a cause and effect relationship and it's placement in the encyclopedia would be all that was ever needed for the entire population to start taking advantage of an increase in precipitation from time to time. The things that humanity has worked out wind up in encyclopedia, and is there, available to those who seek it out. Once the knowledge involved in this presentation is present in encyclopedia, the likelihood of hurricanes developing at all begins almost immediately to decrease, and the same solution that suffices in the marketplace, supply and demand, will stabilize water supplies and lead to higher precipitation amounts over land, and less in the oceans, exactly what is needed.

As economics writers have long argued, simple human interest in self preservation will solve most economic problems, and it appears that could carry over into meteorological events as well. The occasional abuse of this knowledge would be safeguarded against, once all who wish to learn this may. Time to dispense with the take the weather as it comes mentality,no diety is paying our weather any attention. I'm sure that among the youth of the world, the challenge of making the Earth a much more habitable biosphere will appeal to many.

Yes, take the day off! And don't clog up the streets!

I don't really know if this process in nature was ever noticed by anyone besides Tesla himself. Maybe there has never been a deliberate concealment. That would be in keeping with the long time it took mankind to recognize electricity in the first place. Nevertheless, it has been at least discussed by some meteorologists, as evidenced by the one paragraph from our 1980 book with part of the title Weather Modification. It could be that an honest mistake was made previous to 1980, and now with the current theories abounding about dark matter and dark energy it is an idea that is screaming for a closer look. The scientific mind appears to have concluded thus; Hatfield burns some copper compounds, three days later it rains. The atmosphere is comprised of electrically neutral particles, therefore the copper compounds had no effect on the atmospheric pressure, cloud development, and precipitation. The rain was a coincident, and Hatfield was a charlatan. And then we repeat that 3 or 4 times, each time with the same result. It rained, Hatfield got paid, but it was a coincidence again.

Let's return to one of the first sentences of this book, the stance of the meteorological community toward weather modification. In light of what has been discussed, it could be argued there are five or six distinct methods that we know of that can change the weather. Electromagnetic pulse, a plume of conductive compounds sent airborne from a furnace, chem-trails, electromagnets to attract or repel water molecules, and strategically placed nearly pure conductive or non-conductive metals of some size. We could have the Chinese for a sixth, with their attempts to stop rain by firing bombs with water absorbing chemicals at clouds during the 2008 Olympics, though whether or not what they did was effective is unknown to me.

Five or six distinct weather modifying methods conflicts considerably with the notion that only cloud seeding is known to help a little to generate precipitation. There are many ways to skin a cat! As to which method works best and is most cost effective, the metal placements in strategic locations wins. Not only does copper in a high location do what is claimed in this book, it does it so well that the likelihood of rain within 50 miles of such a placement of sufficient size is nearly 100% after 72 hours of the copper sitting there. 100 square miles is blanketed well, areas farther away less and less as distance increases from the copper. Further east of the location of the copper the storm could continue with a life of its own, and usually does. Places along the western coast of the United States have numerous places where the mountains rise right near the coast. Placing copper in one of those locations could see rain daily, with the mountains breaking up the storm as it moves inland.

Knowledge for knowledge's sake is an important part of living for a lot of people. Many non-professional people have a keen interest in science. They specialize in some other area in our division of labor society. They expect the findings of the professionals occupied with the sciences to be eventually reported to them. A shipment doesn't just not arrive from somewhere. Why shouldn't what has been discovered be reported to them? After all, 110 years is a long time, from the first copper meets atmosphere events, more than enough for any patent to have expired long ago. The polar nature of the water molecule should be well known to any scientist concerned with the atmosphere or the environment.

A thorough examination of what might befall the world should this discovery be conveyed to the general public should consider just what the general public is capable of. A profound discovery such as this would generate a lot of interest. Does that mean that x number of people with pick-up trucks are going to try this for themselves? I would guess it means that more people would buy barometers and watch the weather channel more often. It also means that any unusual fall or rise in barometric pressure that has not been planned by the meteorological community will be instantly reported by hundreds of home bodies,perhaps thousands, people who for whatever reason spend a lot of time at home idle, or doing housework, and able to watch television.

She says her barometer fell

The acceptance of our mastery over the weather and the excellent weather that we all begin to experience will eventually sink in, and after a few years, the number of people watching the weather channel because of the novelty of that new discovery will begin to tail off, and probably continue to tail off so dramatically that eventually it could disappear completely, with regular news channels giving tentative weather plans.

The world wide meteorological community would all see that world wide participation would be most conducive to peaceful weather, since decentralizing moisture from the oceans by using numerous copper placements around the world would prevent huge storms from developing.Who will take care of the polar regions will ultimately be decided, and plain ordinary rain could get more frequent. If that is all that happens and everyone knows when it is going to happen, it becomes commonplace; some young people may opt to enter the meteorological field, most people will completely lose interest once singular or calamitous weather events hardly ever happen.

As time goes on,and this type of weather modification activity increases, which may be happening already, it will become increasingly more imperative that all areas of the globe be aware that they may experience a shortage of water. In spring and summer of 2009 there was a drought in Northern China; a drought in Australia with severe brush fires; various provinces in India did not get a monsoon that year; the United States had two or three droughts according to the Weather Channel; Brazil, of all places, was also experiencing drought and this type of situation can only worsen, if everyone remains ignorant but a few.

Naturally not everyone is going to agree as to how much precipitation should be occurring in the area in which they live. The air and water are public things;

access to natural bodies of water is usually made a part of law, no one can prevent another person from having access to a lake even if they own all the land surrounding the lake. Lakes, oceans, and rivers are not bought and sold; they are part of the public domain by law, just as sidewalks are. No one can buy part of the air and prevent others from breathing it, either.

How is the issue resolved as to how much precipitation should occur in a given area? People who suffer from arthritis hate it when the barometric pressure falls,they experience more pain then, all the joints in their bodies ache, they can barely open a screw on cap because of the pain in their wrists, walking can be torture for arthritic ankles during a period of lower barometric pressure. People who ride motorcycles, too,never seem to want any rain to put a damper on their enjoyment of motorcycle riding. Others need to use a motorcycle or scooter to commute to and from work, and a rainstorm is simply not acceptable to them.

Water is a necessity of all living things, and some precipitation must occur if life is to continue at all. It looks as though some people are going to be inconvenienced from time to time to facilitate the continuance of living things on land. Having accurate advanced forecasts would help immensely. At least those who are the most inconvenienced will be able to prepare for rain on schedule. Rain is going to happen, the hope now is it won't be too much flooding before sensible reining in of this capacity to make rain begins to happen.

Economically, the allocation of scarce resources changes continually as circumstances change.Rebuilding structures that have been destroyed by weather may provide jobs, but as far as improving the productivity of labor and increasing the wealth of the citizens, it would appear that retaining those structures that are already built and hence permitting the allocation of resources to new developments would accomplish much more.

You were right about the umbrella. Momentum renders it useless.

A dollar figure would be hard to arrive at when one considers how many structures have been destroyed by weather or forest fires over the past century, and the labor and materials needed for rebuilding these have been withdrawn from other areas. A long period of time with little severe weather anywhere on the globe, along with very few forest fires could usher in an age of prosperity never before seen, along with bumper crops year after year. The earthquakes in Haiti, Chile, New Zealand and Japan are disasters little can be done to prevent. Preventing those disasters that could conceivably be prevented could bring some relief from the overwhelming tide of natural disasters.

A look at the Sahara Desert and the Amazon Rain Forest reveal some interesting differences in metallic abundances between the two areas. The Amazon sits just east of the Andes Mountains, where 70% of the world's copper ore can be found. Copper ore is only around one percent copper, in contrast to copper refrigerator tubing, which is more than 99% copper. A number of bits of evidence point to the Sahara Desert having more than the usual amounts of lead at or near the surface of the land.

The Ancient Egyptians used cosmetic face makeup countless centuries ago, mostly consisting of lead based paints. Mummies unearthed have been examined and been found to possess high concentrations of lead in the bones. It has even been speculated that a fair proportion of these people died of lead poisoning, or at least had their lives shortened by excessive lead accumulating in their bodies.

Drinking from earthenware vessels that were made of clay containing excessive amounts of lead is thought to be the cause of the accumulations in the bones of these mummies, since the lead would leach out of the clay into the liquids being drank, and hence into the bodies of the persons drinking the liquids. The reason lead is more abundant there is because that land area has some of the Earth's oldest crust, that solidified originally 4.5 billion years ago, and has never been torn asunder by volcanoes. It has been covered over by lighter elements, sand.

Between the Amazon and the Sahara we find the Sargasso Sea, the place where ancient mariners were becalmed for weeks on end, and since wind is what the vessels they were sailing on depended, this was a most difficult part of the ocean for the ancient mariners to traverse. Based on the Sahara resting on an ancient crust with a lot of lead, and the Andes Mountains being the largest volcanic mountain range with the most copper anywhere, wouldn't that cause lower barometric pressure on the west side of the Sargasso Sea, and higher barometric pressure on the east side of the Sargasso Sea? Surely that is why the prevailing westerlies no longer prevail there, as often as not.

A meteorology professor replied to an e-mail I sent to him (this was basically my only response to over 100 e-mails to addresses obtained from college meteorology departments) that I obviously didn't know anything about the science of meteorology in general or the weather patterns of the Southern Atlantic in particular, and that everyone knew that the idea I was promoting was a croc. I didn't put the rain forest next to the copper (the World's largest Rain Forest) or the largest Desert in the world on top of an unusually large amount of lead. There is no other ocean on the planet with doldrums as intense as the southern Atlantic Ocean.

The laws of physics dictates what occurs, if the evidence supports it, it must be true. That is the difficulty with weather phenomenon, at present, since this discovery could be easily overlooked and attributed to mere coincidence, starting with Tesla and his experiments in 1899. There have been far too many "coincidences" already to justify rejecting the particular concept that the weather can be easily controlled with a little time, patience and perseverance, in addition to having the truth of the matter on record.

The ivory tower stance that is so prevalent in the scientific community seems to be operative here; the possibility of abuse of this feature of nature by just about anyone justifies keeping over 6 billion people in ignorance. Dangerous or not, the truth should be known.

The means to build an atomic bomb, for example, is not easy to find. Manuals describing how to do this are not available to the general public. The difference between building a thermonuclear device and simply putting some copper or lead in a high location is quite striking; one is a very complicated and difficult undertaking requiring some rather rare and restricted items, along with the way to assemble them properly, and the other is remarkably simple. It is this simplicity and the likelihood that abuse could be prevented that indicate it would be wiser to permit this discovery to become available to the general public. The simplicity part is also a sure fire guarantee that others could make the same discovery just by noticing the differences in metallic abundances and climate existing between the Amazon and the Sahara, even children.

A nuclear bomb can only be used for destruction or the prevention of war. The awful effects of nuclear war has prevented all out wars from being fought. The discovery of a process to control the weather would have countless peaceful purposes. World population tops 6 billion, global warming becomes an ever growing problem and the number of people trying to discover a means to counter rising temperatures, melting ice caps and rising ocean levels among other environmental problems is now in the thousands, perhaps even a million or two. Should we just tell these million or two people to stop trying to solve these problems since whatever solution they come up with won't be used anyway, since whatever it is, if it works, it will work in excess as well, and be unacceptable in the ivory tower?

If it is possible for human beings to take charge of some aspects of nature and control what it does,do you suppose that will ever happen? Volcanoes will still erupt, earthquakes will still happen.The possibility that the Earth could be hit by a large asteroid or a cosmic beam one day is still real,but horrendous weather could have disappeared about 80 years ago, and should certainly be gone now. Most news on a daily basis is bad. Just by starting to avert disasters with the weather could improve the mood of anyone reading the news. When weather disasters begin to decrease, humanity will see that it has a better alternative.

Once I got a computer,I e-mailed a lot of meteorology students and professors when I found e-mail addresses on College and University Meteorology Department web sites. The east coast has a lot of universities. I wonder if that has anything to do with the rather cool, wet summer there in 2010, and increased precipitation over the past 10 years or so.Of the 103 nuclear power plants in the United States,about 60 or more are near the east coast also.That makes the wetter weather there all the more unusual. No reply from meteorologists but one, unless you include "Who are you and why are you sending me this stuff?" (from the east coast), until recently, now that this writing has progressed some. Before 2007 my description of weather modification was too brief and mostly ignored. So this book is the only way, really, to get my point across. No science journal will publish non-scientists work, or only very rarely.

When and if this presentation is proven correct, one could argue that it should be required reading for first year high school students. It might teach the youth of the world more than just weather modification or cosmology. They may learn and adapt, and future civilization might no longer see the lethargy and apathy towards issues affecting the environment, government and the general public, which the last three or four generations of human civilization have shown. Would that make us ultimately more like honey bees if we were better able to do things in concert? Or a colony of ants? I think it would show we grew smarter. I also hope it helps the youth of the world see that it is necessary to question everything, and to think outside of the box, as the saying goes.

Authorities are currently at a standstill since it has not been acknowledged that human activity could change the weather as dramatically as I argue that it can.Were what I have been asserting turned out to be the truth and that was successfully determined and placed in encyclopedia, then the wheels of the authorities would be free to spin based on the new information,confirmed by reputable scientists. Otherwise, misadventures with weather will likely only get worse.

Chapter 6. Benefits of the New, Costs of the Old

"Unspoiled Nature Never Looked Better Than
When It Has Been Spoiled By Plenty"
JB

Desalinization plants have been built, are in the planning stages, or are being proposed in a number of locations around the world. The expense of building a desalinization plant for fresh water far exceeds the cost of timely placement of copper near a mountain top. Water in the form of precipitation will do all the irrigating, all that is needed is water channels here and there. Building a huge factory to produce steam, cause it to condense, and then collect the water, drip by drip, at a cost of billions of dollars when better water distribution is available at a cost of one ten millionth of the price of the desalinization plant doesn't make much sense. The profuse quantities of water available through precipitation and the wide distribution of the water make it clear that precipitation making is the way to go. One decent sized storm would provide more water than a desalinization plant could produce in a month, spread out all over a fair sized area. Desalinization plants also will do nothing to mitigate flooding, which is also theoretically possible with the processes of this discovery we have been discussing.

Here we have a textbook example of what happens when pertinent information is left out of the encyclopedia of the world. Suppose, in a possible scenario, a semi arid country builds a desalinization plant. The engineers who designed the desalinization plant didn't have the information contained herein available to them. The government agency who hired the engineers didn't know this either, that is why they chose to build a desalinization plant. If the engineers had known, they could have pointed out to the government agency that a better way of obtaining fresh water had been developed; either way, information needed to be stored and available, and it wasn't. Now every citizen of that country is paying the price for expensive water purification, either through taxes or inflation. Huge malinvestments start adding up. Hopefully no new desalinization plants will be the rule in the not too distant future.

The polar ice caps are melting at an increasing rate, and the only way to counter this possibly devastating inevitability while reducing fossil fuel emissions would be to load the poles with copper during the winter time at the respective poles and hence cause more snow to fall thereby rebuilding the ice caps. There are permanent settlements in both polar regions. Once again, the encyclopedia could play a big role in what those people do in the polar regions. For one, a lot of time they are indoors and I expect most of them read at least some of the time.

Most residents of polar settlements have college degrees, after all, they are there to study the place. One doesn't drill for ice cores without some education. Learning that the polar habitat they occupy could be in a position to help increase snow amounts on the ice cap from reading about it in an encyclopedia would have these bored prisoners of the ice volunteering to position copper somewhere near the camp they live in. Any break from the ordinary would be welcome. Having to dig out from snowstorms there is already routine.

The increased runoff from land would be beneficial to the marine organisms on the planet. It seems almost all the iron in the sea comes from precipitation runoff from land. Coastal areas are important to the overall health of all marine organisms. The larger creatures that cross entire oceans are dependent upon coastal prey for iron intake, which is probably as essential to marine creatures as it is to land animals, or nearly so. Rivers like the Colorado, that currently are almost completely used up by the time the water reaches the mouth of the river, are not helping to improve fishing worldwide, or the health of the creatures living in the oceans.

The tiny trickle of water that reaches the Sea of Cortez at the mouth of the Colorado doesn't do much to enrich that coastal area with iron. My hope is that once this idea becomes accepted as fact and made proper use of, the trickle coming from the Colorado would eventually be a more robust runoff, and coupled with other means of generating electricity besides building dams could eventually make possible the tearing down of some of the dams already built, returning some river systems to their natural state. Perhaps as dams wear out, and repairing the dam versus adopting new technologies is weighed carefully, one by one the dams will come down. One would certainly hope that in the future humanity will see the necessity of keeping water flowing from land to sea.

Plants and animals in the wild, some near extinction, would all begin to have an easier time if water were more wisely distributed around the world. When one looks at what mankind has done to feed the growing populations of humans over the centuries since the Americas were discovered, it seems the least we could do to help creatures in the wild. The North American Bison nearly went extinct. The dodo bird and passenger pigeon are extinct along with a number of less well known creatures and plants. Amphibians have suffered over 100 extinctions over the past century.

The entire downward spiral of living things in the wild that has been going on since man began encroaching upon territory that was once occupied only by wild life could be reversed. It surely adds to one's enjoyment of nature if one actually sees wild animals and plants in great profusion if one were to travel to a national park. Humans happening to be micro managing the weather to provide water for the living things there isn't going to detract from the beauty of the place. On the contrary, it would help a great deal in helping wild things grow, and one's chances of seeing some creature or plant that is rarely seen go up.

No trace of what is taking place with the weather would appear as an unwelcome change in the wild. The modification of the weather involves no building of fences or any other thing to impede the movement of animals in the wild. Something placed somewhere in the area periodically poses no danger to the living things near it. The animals walk around it.

If, for example, the desert southwest of the United States embarked upon a weather modification program designed to provide a decent sized storm about every ten days during spring and summer, and perhaps once a week in fall and winter, the populations of desert rabbits and tortoises, various birds, and almost all the plants in the area would begin to increase. After a few decades it might even be possible to hunt rabbits in the southwest. Mountain lions would also see an increase in populations with fast reproducing animals like desert rabbits in greater abundance.

After a while, driving through the area could prove a lot more interesting with more plants and animals around. Eventually, problems will inevitably arise when some wildlife encroach upon human settlements, but that is something that happens already. Moose walk through Alaskan towns routinely, and deer, bear and lions occasionally wander into towns all over the United States. Some of the countries where primates live have the problem of dealing with thieving primates everyday.

Incursions into human areas by wildlife, then, could increase with increasing populations of wild animals, but these incursions might possibly decrease because the animals are finding enough to eat in the wild. Bears don't visit human habitations for company. They are there searching for something to eat, perhaps because the food in the wild is insufficient at the moment.

The question of what becomes of wilderness areas and how weather modification activities will impact upon populations of wild animals and plants will just have to wait until enough time has gone by to make some kind of assessment. With the use of such devices as described herein,there is no doubt but that increased populations of living things in the wild will result.

Humans have always forced their way into wilderness areas and built settlements where wildlife wasn't far away and tended to encroach, so nothing is really going to change substantially. Land owners themselves are faced with the decision to build a fence or wall,bring in cats to keep out mice, dogs to fend off wolves and mountain lions, etc. A settlement sure seems more appropriate where there are a lot of plants and animals, and available fresh water, as contrasted to a desert area.

That macaque stole the pie from the windowsill!

Whether this increase in populations of animals in the

wild causes further problems for humankind, that is entirely possible. Perhaps we will see more locust plagues, greater trouble with Africanized honey bees, more problems with nesting birds, more outbreaks of mosquito -borne illnesses, etc. None of the possible problems that wild populations could present would be problems humans have never seen before. But then, wild populations have never had weather conditions optimized for them before.

Perhaps some creature will become way too numerous and pose such a problem that the only way to stop this problem creature would be to stop any and all precipitation in a geographical area until the creature dies off. Fortunately, the theory gives us a way to do that. Any runaway population explosion of some disease vector animal or whatever kind of problem plant or animal will at least start out locally, and if local health authorities are vigilant, could be stopped before spreading around the world.

Life is by necessity opportunistic. A Burmese Python in Florida doesn't bemoan the loss of its ancestral home, it welcomes the new types of food it can eat in the Everglades. So many species have moved to new habitats recently because of human activity that it will be impossible to eradicate the new inhabitants.Prior to human beings acquiring exotic pets,migrations occurred in the natural world also. Plants and animals all try to expand the range of the species. Whether a change of habitat happens naturally or through human carelessness, the universe isn't going to make a distinction. The entire Earth is a possible habitat for every living organism. Some can only live on land, others in the sea, but each creature has a shot at establishing itself in other locations, and eventually circling the globe, by land or sea. Throw flying creatures in with land creatures for the sake of simplicity.

Increasing precipitation worldwide means changing conditions for cacti. Cacti do not find water inimical. They will develop a means of employing the excess water. Cacti will evolve. We will be able to witness evolution. Eventually, new plant material will start to make an appearance on cacti, like leaves, and some of the water storage cells that are no longer useful will start dying off. A change would be triggered in the DNA of the plant by the change in precipitation.

The possibility that cacti could go extinct is real, with the exception of those in controlled environments such as a greenhouse. They could be overrun by invading plant species that grow much faster to the extent they could be choked out of existence. Desert locations will want to proceed slowly and gradually with increased precipitation to allow local fauna to acclimatize itself to the change, and to slow invasive species. There are some who would argue that climate change should be allowed to proceed, and let survival of the fittest determine whether cacti, a plant that developed in arid land areas equipped with adaptations allowing it to survive prolonged drought, will survive wetter climates in the future.

The whole idea of interfering with nature will meet with opposition, though only benefits could be seen as a result. Even global warming fanatics will not welcome this theory even though it introduces an easy solution to that problem. Nature, however it comes, is for some mystical reason better than having perfect growing seasons, an absence of damaging weather, and higher polar ice caps. Optimizing the weather would cause populations in the wild to double and triple but is still no substitute for the real thing, that which happens if man does nothing.

Barometric pressure is always higher over land than over sea, but we are exacerbating that situation with our civilization. One must realize that human activity modifies the weather in more subtle ways already, not the kind of deliberate storm making or stopping that is being discussed here but simply higher barometric pressure over land. Mother Nature, what one refers to when the course of natural atmospheric events culminates in rain, drought, flooding, whatever, is only happenstance, and in this day and age, happenstance will lead to drought more often than not.

300 years ago, there was not a single airplane crossing the sky; nary a motor vehicle or for that matter an asphalt road. All the hundreds of airplane flights, millions of motor vehicles will bring the air pressure up even more so over land than previously. Throw in all the additional asphalt and concrete, houses painted with lead based paints, common until not long ago, and nuclear power plants on land, all with thick lead walls, and the air pressure over land can become so high that rain might never come by itself.

The thing to do, then, is what? Doing nothing and letting "Mother Nature" take care of things isn't going to work. Human activity has already tipped the scales against precipitation occurring on land, unless it's a hurricane. We are ultimately responsible with our civilization for the higher barometric pressure over land, wouldn't it make sense to lower the barometric pressure, produce clouds and precipitation, by use of another process developed by our civilization? I didn't develop this process, it appears to have been around before I was born. The important thing to establish isn't when this first got noticed, but whether it really does what I claim. If correct, then it is a process that should be available to anyone who is curious about it, a process no one owns but one of nature that humans have now grasped.

The increased use of this device would be a great help in reducing air pollution. Major cities with smog problems like Mexico City and Los Angeles could plan on more frequent showers to clean the air more often. Most of the carbon dioxide would end up in the oceans eventually, but at least the atmosphere gets cleansed of it and other air borne particles. People with asthma would have an easier time.

Precipitation is about the only way CO_2 and other pollutants can be reduced, once the pollutants are airborne, except for increasing amounts of photosynthesis. A recent study suggests that rising carbon dioxide levels in the atmosphere which are absorbed by the oceans in vast quantities may make the oceans too acidic and cause, perhaps, mass extinction of a sizeable percentage of marine life. A lot of the coral reefs are already struggling, and this could worsen.

The larger populations and sizes of plants that would result from increased precipitation removes a greater amount of CO_2 through photosynthesis. Increased precipitation over land areas removes water from the oceans and fills rivers, lakes and aquifers. Even allowing desert dwellers to use water without fear of running out would help keep ocean levels from rising to some small extent. Currently in Las Vegas, were I to wash my car in my driveway, I could be fined for water wasting. Convince me that the law has this correct.

There has been some research on funneling air into CO_2 scrubbers that remove the CO_2 that comes in and release air free of it, but whether that idea will be developed on a large scale remains to be seen. But at least we are beginning to see some scientists looking at things that could be done on a large scale, covering planet wide activities.

We may see the day when several dozen huge CO_2 scrubbers are put into use around the world, along with a modern approach to the weather by meteorologists. Every additional rain storm, or even cloud over land slows the global warming problems, postponing doomsday a little longer, until perhaps we have found some feasible solution to the problem.

Weather forecasting could become a planning event rather than a guessing event. Should a rain event be planned, there would be no reason not to tell the public about it; hence people could plan accordingly. Boating outings could be planned for the days between rain events. This could be a revolution in advanced planning.

After twenty years or so, organizations like professional baseball might decide to schedule games and days off in sync with the weather forecast, which by that time would probably be 100% accurate. Fewer postponements due to unforeseen weather will likely be the rule. Unforeseen weather just will not exist.

Forest fires have been fought in some instances for weeks, with fatalities occurring among the fire fighters. The National Forest Service continually monitors all forested areas, and are usually aware of a forest fire within an hour of its origination. Given the wind conditions and the likelihood of the fire becoming difficult to contain, copper could be in place beginning its seventy-two hour storm incubation within three or four hours of a possibly dangerous fire. The fire fighters could concentrate on containment until the storm arrives. Earlier, it was pointed out that wind could be minimized by creating a path of least resistance that countermanded the prevailing westerlies by an adjustable angle. Placing copper somewhere high up to produce precipitation to put out a forest fire, those performing this duty should try to position the copper in such a way as to minimize the wind in the area as well, and that could help significantly in reducing how far the fire spreads before precipitation arrives.

With use of this process by properly cautious meteorologists few areas would be prone to forest fires because most places would be receiving enough precipitation to keep the vegetation moist, and not likely to burn out of control, although in the case of arson or accident a fire would still be possible between planned rain events.

Windmill farms could use horizontally placed copper to augment the wind and increase the yield of the windmills. After a few decades, certain areas will have designations assigned to them based on what type of economic activity is transpiring there. Places where windmill farms are in abundance will have a windy designation. Those localities concentrating on growing litchi trees, or other wind sensitive plants would have a calm designation. Right now, Southern California has a large windmill farm, and a lot of agriculture at the same time, not far from the windmills. How that will work out will take time. A lot of plants can thrive in windy climates, especially root type edibles like potatoes, onions and carrots. After a long time the windmills might get moved out to sea.

Shipping could see a marked improvement worldwide. Fewer ships would be lost at sea if the shipping lanes could be freed of storms. A ship laden with a ton or two of lead on its deck could patrol the boundaries of shipping lanes and intervene between approaching storms, causing them to either change direction or weaken in intensity, or both.

The ocean is basically flat, and barometric pressure is on average much lower out over the oceans than on land, where the uneven terrain, the lower percentage of water evaporating skyward and the more extreme temperature changes between night and day all contribute to rising barometric pressure. Nowadays one can add human activities to factors raising air pressure over land.Diverting the miniscule, free floating molecules in our atmosphere toward land is possible, and together with the diversion of water vapor towards the poles would cripple the large storms we see now, and make ocean travel safer in and of itself, even without a vessel carrying lead on patrol. After a considerable time, with all the major land areas intent upon an adequate supply of water, the oceans could wind up flat as glass and atmospherically dry as a bone for the most part. The only time precipitation would be found in the oceans is when some islands chose to have some.

Some land areas have extremely uneven terrain

Rising ocean levels is the prediction from global warming alarmists. All relevant information has not been accumulated, or at least not enough to know with certainty what the future holds in this regard. There could possibly be nothing to fear; our puny efforts are overwhelmed by the inexorable changes that the Earth goes through, and temperatures could start declining on their own. Based on the NASA radio waves detected in 2006, a realistic estimate of the future in this regard is that global warming will probably continue, since we have been or are now being flooded by a new source of hydrogen. Another big bang in the neighborhood of the known universe would do that. New hydrogen in increased amounts means greater output by the sun, more water on our planet, and melting ice caps.

Global warming is a good thing for people in cold climates, and if it eventually happens, if it happens slowly enough the end result before some ice age or another would probably be good. If the Earth is on a warming trend, with or without humans burning fossil fuels, there is little we will be able to do about it but reap the benefits of longer growing seasons and more arable land. Efforts to maximize polar ice caps and add more water to the land will slow the warming trend down some, perhaps enough to make the transition easier.

For all we know, we have our hands on the Earth's thermostat and just don't realize it. Years will have to elapse with this solution working that we now have to see if world wide temperatures aren't more easily controlled. With world wide use the cloud cover difference over land would be enormous. The increased plant growth that results from plentiful precipitation worldwide ramps up photosynthesis to a much higher level, reducing CO_2. The polar ice caps could be enormous after a decade of increased snow.

We can also throw in the cost of continued searching for dark matter and dark energy, when computing the costs of letting this idea slip through our fingers, if it ultimately is proven that more hydrogen is coming into existence near the path of least resistance created by the existence of a strategically placed row, or rows, of copper tubing.

One experiment to confirm or deny this could consist of an experiment with copper such as has been described, with random air samples taken in the immediate vicinity after 72 hours, followed a week after removal of the copper with an experiment with thin lead sheets, all placed the long way north-south so the entire face confronts the prevailing westerlies, with identical air samples taken after 72 hours also. How many hydrogen atoms from each series of samples of the same size could be determined, and if the samples from the copper experiment showed a higher percent concentration of hydrogen, the missing dark matter and energy are found.

We wouldn't have exactly found the missing particle, but we would know it would accumulate in greater concentrations along a path of least resistance and that it is there combining with others of it's kind to become hydrogen. That may be the only way we can confirm the existence of a primordial speck, if it eludes detection as itself. They do rather neatly fit into an explanation for dark matter and dark energy. Newer equipment designed to capture a hydrogen atom appearing as though from nowhere might eventually be created. However many millions of dollars and scientists, buildings, equipment, etc., are expended in the search for dark matter and dark energy could be expended on more promising pursuits if these experiments are successful.

Another experiment to confirm or deny hydrogen coming into being would be simply placing a vacuum chamber at ground zero of an experiment with copper in an elevated location. If the chamber was empty, or as nearly so as humanly possible at the start of the experiment, sat there seventy two hours and some hydrogen was discovered in the container upon later examination, that would be pretty conclusive evidence that things small enough to pass through the walls of the container did so, and then became hydrogen. The primordial specks could get into the chamber, but if some combine to become hydrogen there, the newly condensed hydrogen is trapped.

The size of the vacuum chamber could be a limiting factor. If the chamber is too small, the precursors to hydrogen might not have enough space to condense into hydrogen between the walls of the container. The container being almost completely devoid of matter might also influence how readily hydrogen condenses. Depending on several unknowns, the outcome of the experiment could be that there is no hydrogen within the vacuum chamber after 72 hours. That would not conclusively prove that no hydrogen is coming into existence along a path of least resistance; if hydrogen was found it would prove that it was. A larger vacuum chamber might yield a different result for all we know. Pinning down a tiny hydrogen atom coming into existence isn't going to be easy.

Finding any hydrogen at all where it couldn't be unless it just popped into existence would give us a considerable insight. The vacuum chamber could have a set amount of hydrogen in it rather than a complete vacuum, in the event that hydrogen only comes into existence within a paired hydrogen isotope serving as a template. With the vacuum chamber empty the same experiment would yield a clue if the experiment with the chamber with some hydrogen in it proved to have increased amounts of hydrogen after some time,and the completely empty chamber had none. We would then know that hydrogen was coming from hydrogen already existing, meaning it started slowly in the universe and could possibly be still picking up speed as more and more templates come into existence. If both chambers contained additional hydrogen after time, we would be fairly sure that hydrogen was coming into existence spontaneously in the universe. If that were the case then hydrogen production in the universe could have begun quickly, and slowly tailed off. Both chambers having no additional hydrogen is also a possibility. That would not preclude precursors to hydrogen in stars from engaging in hydrogen reproduction, possibly at a considerable rate.

Knowing whether hydrogen production or reproduction is on going or not and to what extent would help enormously. The scenario I suggested, that the ejecta from the Big Bang couldn't possibly know how to do anything after they are blasted into space from the explosion, it took some time for one hydrogen atom to develop or some paired hydrogen isotopes to drift into the early known universe from elsewhere, and that hydrogen atoms reproduce within already existing hydrogen atom pairs that serve as templates for information transfer completely reverses the order of hydrogen production in the known universe.

The current theory is at all happened very quickly and is finished now. That is the scientific conclusion, though it certainly begs the question of how anyone could possibly know what took place after the plasma of the early universe cooled. If the possibility I suggest is the true one, the current state of the known universe would be the time with the greatest number of templates to facilitate more hydrogen production, and tomorrow there will be still more, and so on. Quite a difference between the statement that hydrogen production in the known universe ended long ago and the suggestion that it could possibly still be on the increase.

A more likely situation would be hydrogen beginning to reproduce slowly, and the process speeding up for 5 or 6 billion years, then leveling off and slowly winding down. The known universe still contains large quantities of dark matter and dark energy, so one would have to assume that,though hydrogen production may be tailing off some, it is still ongoing.

If hydrogen begets more hydrogen by some spontaneous process involving hydrogen pairs serving as templates, once stars begin to ignite and burn through nuclear fusion, the process of hydrogen production slows to some extent because existing templates are being reduced as hydrogen fuses into helium, and supernova produce more elements higher up the table of elements all of which began as hydrogen but are now some heavier element.

Whatever all the processes are that result in the heavier elements than hydrogen coming into existence, they would all take away possible templates for more hydrogen production. The greater the diameter of the totality of things we know as the known universe, as billions of years pass, the more diffuse the primordial specks would become. Hydrogen is the most abundant element now, and will probably hold on to a high percentage of known matter for a very long time. For as many stars as there are now yet still hydrogen is most abundant and dark matter and dark energy are huge percentages of all in the known universe, I would guess my theory fits the known facts better. Most hydrogen production would be going on within stars, where gravitational attraction is stronger than here, increasing the concentrations of primordial specks, and H_2 isotopes are in far greater abundance.

Now that stars and galaxies are widespread, stars have all probably established a steady rate of hydrogen production that is likely to continue for another 10 billion years or more. Some stars will explode, others will ignite for the first time, and the process of hydrogen growth will continue until there are no longer sufficient concentrations of primordial specks to continue the process. Black holes are going to gain a substantial percent of the primordial specks over the next 10 billion years, but in the meantime, hydrogen will have continued to reproduce itself. Eventually the percentage of matter that is detectable could increase to as much as 40% of the entire matter and energy in the known universe. At that point, stars would begin to run out of fuel, and the entire known universe would begin to go dark. How long that will take is probably at least as long again as the known universe has already existed, possibly longer.

We also have people searching for a solution to these problems with the weather that we are contending have already been more or less solved, just in need of further experimentation. How much time, energy, capital and equipment that is being thrown into such pursuits is probably considerable. The burden of proof to me falls to the scientific community,or at least to the astrophysical community, to ascertain the results of the experiments just described, whether hydrogen is more abundant in the copper part of the experiment and less abundant in the lead part of the experiment, and whether hydrogen will accumulate in a vacuum chamber empty, or with hydrogen isotopes within the chamber to serve as template pairs. It behooves them to conduct such experiments.

The costs of living with nature as it comes will rise, there will also be the stunted intellectual growth of children to consider. Concealing pertinent truths from children isn't wise. The more we are free to learn, the more we can prosper thereby. Worse truths we live with every day, like guns. The younger and sooner one learns something the better able to adapt and adjust to the new information one is. Contrast the skills of the young when using a computer in this day and age to the skills of the old in that regard. There are still people alive from a time when no computer existed. Prior to 1970, there were computers, but not really in the home yet. Quite a lot of old dogs learning new tricks, slowly.

Since the new process described in this book isn't really new, it could be as old as the Roman Empire, when do the encyclopedia begin to contain it? It seems that without official confirmation by those scientists with degrees in the appropriate sciences, a discovery this breathtaking just founders, with only a few people aware of it, for decades more. A petition at the end of the book, perhaps. When children grow up in a world where this discovery is properly dealt with, and is available in encyclopedia, the bright and resourceful will never miss a step integrating that new information.

Just because the last three or four generations haven't pieced all these new factors together about dark matter and dark energy and tied that together with meteorological events, and added that to already existing knowledge of the polar nature of the water molecule, doesn't mean it will never happen. Better the young get as much as the previous generations can provide to better equip them with knowledge of how to deal with things.

Besides the worsening forecast for upcoming decades, there is still the truth, and if it isn't addressed soon, anyone who cares to try a strategic placement of copper or lead of their own won't be held accountable for anything in the event that they are seen doing this experiment. They would be forced off public land. Probably that would be all.

Mischief such as described might not all be mischief, if no one else seems to be trying to do this, why don't I? So everybody and his brother would think, during times of drought. That kind of thing, if left unchecked, could prove disastrous if too many people got involved. Therefore, I think the grim truth must soon be faced for the world to see better times rather than worse. The only way it can succeed is by more and more people learning of it.

Eventually it will break above ground and sprout beautifully, and take its place amongst other great discoveries. It will become an accepted part of the routine of living. Nearly everyone will be on the lookout for any unplanned changes in barometric pressure. The absence of severe storms and an easy way to distribute water to every land mass on the planet would make it possible to prosper as never before.

Plentiful water whenever needed is certainly going to improve the economic prospects of just about everyone. That economic gain wouldn't always be easy to see, as the years pass, and the Earth becomes a much more habitable bio-sphere for all its inhabitants. Regrowing entire forests usually starts with a dedicated group who walk along together planting seedlings. These environ-mentally friendly volunteers could take that a step further and assure these forests flourish with abundant precipitation.

As these new processes slowly gain worldwide acceptance and use, mountainous terrain localities will need a few more water channels than more gradually sloping terrain, and a lot of the mountainous areas of the world will have aquaduct systems resembling those of ancient Roman and Mayan cities after a few decades.

Chapter 7. Looking Ahead

"The Happy Middle Ground Is More Fertile"
JB

I think that any country that begins a weather modification program using the process we have been discussing should conduct sweeps in a helicopter with a barometer in any area of that country that has seen a lot or very little rainfall in the past decade or so for man made metal deposits. Copper or lead may have been placed in remote locations and abandoned, and these lead or copper placements would need to be found. Indeed, the desert southwest of the United States has remained unusually dry for quite a while.

Arizona, Nevada, and Southern California are the favorite retirement locales of a number of people. It wouldn't surprise me to find that the desert southwest may just be one place where lead has been positioned and left there by some arthritic old codger, who is probably long dead now. This person may have just been someone who, like myself, put two and two together, but decided to do something entirely different, like move to the desert southwest and make sure it stays dry.

For instance, the heaviest rainfall total recorded in the U.S. in a 24 hour period happened in Holt, Missouri back in 1947. I happened to drive by there a few years ago, and noticed a railroad track running through the area. Suppose a huge shipment of copper on open flat bed rail cars happened to be sitting on a sidetrack there for a few days, and then the intense deluge that came along happened. Is it possible someone noticed that? I realize that this is only speculation, but if such a process in nature really exists there is no preventing someone from making the causal connection when they witness copper in large quantities and a storm that followed three days after, or to logically extrapolate what would be likely to happen if lead in quantities were used instead.

I'm sure that once the responsibility sinks in around the meteorological community, the proper steps will be taken ensuring that rainfall from the storms created almost always falls within the one-half to one inch range, and that is all that usually happens, with the exception of some thunder and lightning. It is difficult to ascertain how often intervention would be necessary; once a shortage begins to occur it would certainly behoove the peoples of that area,or the meteorologists entrusted with that responsibility to ensure that crops do not shrivel and die, wells run dry, etc.

If such shortages do not occur,well, so be it. Action in the form of generating storms would not be necessary. How many firemen sit around firehouses doing nothing because there are no fires to put out? They are a safety blanket, they are there in the event of an emergency. The time has come to be prepared for weather emergencies in a similar fashion. It would not be an emergency requiring people on immediate standby, suited up and ready to go like firefighters, but rather a group of people ready to suspend what they are ordinarily doing for three or four days, place some copper tubing or lead sheets in an appropriate location, wait to see the results, and eventually dismantle the copper, or lead, as the case may be, and return to their usual lives. Semi retired people could do this. In the event of a forest fire, where the heat from the fire and all the dust and smoke may interfere with cloud development, it may just be necessary to position more copper than is usually necessary to produce some rain.

The extraordinary minuteness of the as much as 96% of the universe that is dark matter and dark energy, and the astounding abundance of these likely charged particles means that they are much easier to influence than the stable isotopes of nitrogen and oxygen and the other 1% of the known atmosphere and this raises the possibility that soothing music sufficiently loud in a high place might be feasible also,with the sound waves creating a smooth path, a path of least resistance.

Then one couldn't discount the rain dances of North American Indian tribes, or the possibility that human brain waves could be enough to begin a cascade of particles sufficient to bring about lower barometric pressure. All we need do is think the weather we want, and it happens? It would sure be cheaper and easier than moving heavy amounts of metals to high places.

Maybe placing copper somewhere just once on high for three days and informing the public so that local people witness the placement, have a big party on the hillside, and remember the rain that followed and have vivid memories of the copper sitting on that particular hillside, the memory of enough people recalling it once a week could alleviate any water shortages over for the region for a decade. A video of the event could be replayed in a public outdoor theatre once a week.

With groups of people, any number of climbing expeditions could prove interesting if each hiker wore a three pound copper bracelet on each arm. The terrain of the hike wouldn't need to be arduous, just an uphill slope that yields onto the western side of some hill or mountain. There the group could picnic, even camp out for 72 hours, all the bracelets hung in trees around the campsite. An escape plan would need to be thought out. If the campsite suddenly becomes cutoff from the path down by a swollen river from heavy rain, and there is no way down the hill, bury the copper!

A well provisioned safari may one day include a quantity of copper for precipitation making during the safari. Places like the Australian outback and the Gobi desert could be traversed much more easily with a supply of fresh water. Ponds of water being plentiful with occasional rain make the trip livable. Wildlife would accumulate near any water, so hunting could improve.

Once the paradigm shift is made, there will be no turning back. The encyclopedia must convey the truth of what has been discovered in the field of meteorology. Once that has happened, everywhere people will see meteorology in a whole new light. Slowly, the entire economy of the world will begin to grow in a stable world where the flow of goods and services is augmented countless ways by this new discovery.

For example, a village in an impoverished nation sees plentiful rain for a few years. A group decides to open a restaurant in the village, something that has never been done before because of the dry seasons that normally occur. No water, no way to wash dishes in a restaurant. It makes possible any number of things, the water to fill the radiator in a car or truck, one crucial for trade. One could go on indefinitely with examples, especially where the existence of a chance to grow things would lead to further possibilities.

Water is the universal solvent, besides being indispensable to living things. All of the additional things that could be done industrially with a continual supply of water would fill up a long list. Making paper, smelting metals, cooling plastic parts after creation in injection molding machines, the list goes on and on. A country without a single plastic injection molding machine due to impoverishment and dry seasons could begin to see some of those, and, of course, all the thousands of things that can be made with plastic.

The number of well intentioned people will always outnumber those with other intentions. When the entire world is watching just about every weather event the likelihood that all would go well and according to plan is high. Unfortunately, preparedness for misadventures with this type of thing is currently non-existent. Raising awareness about what may really be taking place when disastrous weather strikes needs to begin, and facing facts head on rather than conveniently missing them would help.

Scientific verification lacking, and myself being unable to provide anything more than anecdotal testimony, this process of nature that has been described here may just continue to be completely ignored. I hope this is not the case. We've reviewed the possible causes, the costs and risks, as associated with doing nothing. Now it is up to others beside myself to look over this possibility, hopefully as many people as possible. Since the whole idea of modifying the weather involves an incursion into public things, the public should certainly be the first to be aware of it.

What could go wrong with the environment if the Earth continues to heat up? Rising ocean levels will put many coastal cities underwater, and render them uninhabitable. The cost of relocating millions of people should be added in with the cost of desalinization plants when one compares the cost effectiveness of using strategically placed metal deposits versus ignoring this process in nature we think exists, and continuing as we have.

Land requires precipitation for life to gain hold. No doubt the first land plants and animals left the sea and began to exist on land where precipitation was abundant. To have knowledge of a process that could transform vast areas into places where life thrives shouldn't be ignored on the basis of some obscure finding from over 30 years ago.

Places like the remote canyons in Utah could be seen as possible locations for human habitation in the near future. Building cities where nothing to speak of can grow makes the most sense.Tucson,Phoenix,Las Vegas, should be farmland or wildlife habitat, and eventually could be. The changes that could happen when water can be distributed easily could leave some of the southwestern cities unrecognizable in 50 years. A little blasting, a few roads, and modern ecological cities can spring up in the most inhospitable places. Ecological solutions come easily when one can factor in flowing water and be reasonably sure the flow will not stop. Another Dust Bowl should never happen again. The canyons of Utah, with water and roads, and a little unique engineering could serve as dwelling places comfortably. Flat, valley land would eventually be put to its most economical use, as farmland or wildlife habitat.

Wood could see revived usage as a building material, since new forest area could be planned just about anywhere. My first thoughts about this discovery back in 1980 included the realization that had this process been properly integrated by humanity by 1920, wood would have been put to much more use, and would still be going strong now. Trees are huge plants,and provide habitat for many other living things; being able to grow considerably more of them accomplishes many things. Blanketing the Earth with forests as plentiful as existed hundreds of years ago before the industrial revolution would be the goal. Some of the rarer woods like mahogany could be farmed more easily and be more readily available, at a lower cost.

Price figures on desalinization plants are somewhere around 40 or 50 billion dollars apiece. Add the occasional forest fire that gets out of hand, drought, flooding, hurricanes, tornados, all stronger than ever, and we're broke! The world economy cannot afford not to have this simple solution working. As pointed out earlier, knowledge properly positioned in the encyclopedia is all that would be necessary, human self interest would take care of everything else, including how to handle misadventures with unwanted flooding.

Now that there are two possible candidates in the Cosmos that have been discovered that may well have originated somewhere outside the area of space that is our known universe, the entire Big Bang scenario will be reviewed and revised over time. The notion that the Big Bang began from a singularity, where ALL the matter and energy, time and space also, were squeezed just prior to erupting in the Big Bang, is the first supposition to fall.

Hawkins goes further than that, concluding that the singularity from which all space, time, matter, and energy emerged just came out of nowhere, and that prior to that event, nothing whatsoever existed. Years could pass before an official conclusion is reached as to the existence of things outside the known universe, but reaching that conclusion now and seeing where it leads brings us back to the inescapable conclusion that space couldn't possibly be a fabric if all hadn't been within the singularity, since the singularity couldn't have happened with space, time, and other things outside of the singularity. Discovering our true path through the wilderness will get to be more of a priority once the spurious idea of space as a fabric is thrown out.

Cosmologists will begin writing computer programs dealing with possible cosmic scenarios where there are 100, or a 1000 times more systems of galaxies like our own, with x number of black holes scattered about, scads of empty space, and begin to crunch the numbers until about 52% of the dark energy is contained in systems of galaxies and black holes at varying distances from us. One of these simulations could give us a good estimate of what the entire universe contains. Beyond some incredible distance gravity could be so weak we would never know if there may lie more universe.

The idea that black holes are responsible, in all likelihood, for the Big Bang, and for providing the universe with new energy in general will get more exploration over the next few decades. Some cosmologists might even work out how long until the stars in the known universe burn out, and estimate how long after that before another Big Bang. We can all see these are events that take place over billions and billions of years. Where would the black holes that collided and created our known universe come from? From a dozen or so neighboring systems of galaxies that blossomed, burned out, and were absorbed by black holes over the 500 billion years or more previous to the Big Bang. Once in unknown billions of years things in the universe enter, and eventually depart,a black hole. It is not always the same black hole.

If it turns out to be every 700 billion years, or even in the trillions of years, what real difference would that make? We have the long term itinerary for the entire universe, the exact time frame is unimportant. How much time have the atoms that comprise the things on Earth spent inside black holes, and how much time out stretching their legs, like we are now? Most likely everything in the universe that is matter and energy has spent a far greater time within a black hole than being something thrown out from one. Black holes are home. They roam the universe for eons, growing continually. Right now we are on an extended vacation in the wilderness. It is likely that more than 90% of the time things will be residing in black holes in very tight quarters.

Taking the Primordial Speck Theory to be the paradigm, all forces of nature are accounted for, we are short nothing. We know where the dark matter and dark energy are, and can roughly estimate their distribution throughout the known universe and beyond. We know why galaxies are so unevenly distributed throughout the known universe. What caused the Big Bang also neatly falls into the paradigm, the collision of two black holes, not the entire universe, that occurred in one quadrant of a much larger universe.

Nothing?! Preposterous! The universe can only be eternal

It is also possible to extrapolate conclusions about the remaining universe which we have never seen, since the progression of matter and energy to and from black holes over hundreds of billions of years would always result in systems of galaxies coming about from the collision of black holes that have become overly large, and these would eventually burn out and be reabsorbed by other black holes, a continuous cycle that may never have had a beginning. We have a hypothetical origin for hydrogen in the known universe, quite different from existing theories. To top it all off, we have an explanation for all the anecdotal evidence in the next chapter. What the primordial specks do to the atmosphere along with the water molecules gives the science of meteorology a whole new set of dynamics.

Chapter 8. Anecdotal Evidence

"It Is Patently Clear To
The Most Casual Observer"
JB

A Russian born lady changed her name to Ayn Rand, developed an entire philosophy almost from scratch by herself that was based upon reality and reason, and wrote a number or books, both fiction and non-fiction. Her fiction books had philosophical overtones. Do you suppose it possible that she was trying to point the way as a philosopher because she had some insight into the workings of things that other humans lacked? This discovery we have been examining suggests itself, though Ayn Rand never wrote anything about it, as far as I know. Her best known work was Atlas Shrugged.[9] In that fiction novel there is a discovery, it involves copper and the atmosphere, but there the similarity ends. Or does it?

For Ayn Rand to write a suitable mystery novel with philosophical overtones, poetic license took over, and the details of the discovery in the novel were left out. In the novel, instead of placing copper somewhere on high, and causing precipitation while at the same time capturing lightning and converting it into electrical current, the inventor had discovered a way to capture and convert the ambient static electricity in the atmosphere into electrical current. Not much more than that is provided. It is a long book, with a lot of character development.

It is in the character portrayals that most of the philosophical overtones shone through. The inventor, John Galt, is mysteriously missing, and so is this purported discovery. When the heroine Dagny finally catches up with him and the two are together in his hideout in the mountains of Colorado where the invention provides the electricity for the small village there, there is even then no elaboration as to how the thing works. So without going into too much detail, this novel can be construed as a kind of anecdotal evidence.

Further indications that Ayn Rand, in writing Atlas Shrugged, gave subtle indications about this idea we have been entertaining here, are as follows: Dagny tries to track down John Galt, finds he worked at the Twentieth Century Motor Company somewhere in Wisconsin, visits the now abandoned factory, finds some notes and some copper tubing. When Dagny follows another airplane into the mountains of Colorado, the other airplane carrying the scientist she had hired to solve the problem of how to convert static electricity into electrical current, she loses the other plane in the clouds, and unable to see land, comes to a rough landing precisely in the village where this invention is operative. Earlier in the novel she visits someone in New York, too late to stop him from abandoning his current life and joining John Galt in the hideout in Colorado. John Galt had just left the person Dagny visited. At this time, hard rain was falling in New York. Galt's best friend Francisco was enormously wealthy through family ownership in copper mines in Chile. There is even a part where Francisco deliberately mismanaged the mines and the stock market crash it brought on was played out. There are clouds, rain, copper markets, and copper tubing in the web of intrigue in the book. There is even a nightmare HAARP type of scenario where the crumbling government develops an electromagnetic pulse type weapon, and vaporizes a field of sheep in a demonstration.

Ayn Rand was born Alisa Rosenbaum, a Russian Jew. Her father had a chemists shop that was seized by the state, profoundly influencing the young woman's view of government. That government which governs least, governs best, would be the best way to describe Ayn Rand's view of government. Her view was that capitalism is the ideal form of government,but claimed there has never been a true capitalist society. America, to her, was a sad compromise between capitalism and socialism. She really hated socialism. When 21 years old, she changed her name to Ayn Rand on a ship to America in 1925. Atlas Shrugged was published in 1957.

If what Tesla did and the idea we've been discussing were known by word of mouth in the early twentieth century among intellectuals, maybe what the young Russian lady, recently a college graduate, chose as her name makes sense. Right now I'm sure no old news article will surface from 1922 that mentions anything about this,so we don't really know what was happening back then, whether people talked about the weather in Colorado Springs when Tesla was there. Ayn Rand mentions several times how the early twentieth century in America was a place where common sense was more prevalent than at any other time in human history. She really glorified that shining era,the Age of the Engineer, thought there was no other to compare. Why?

Though absolutely certain that the results that I claim will indeed occur, as for proof, that is less easy to provide. I had never seriously contemplated putting any of this into writing. It is distressing to me to realize now that I should have been taking pictures from the very beginning, even saving news clippings when I wrote somewhere and precipitation happened. August of 2007 was when I began writing what became the first edition on February 22, 2010.

In 1980, when I happened to first perceive the effects that I have been describing, a major drought was ongoing in the Southern U. S., and I wrote a letter to the Society of Separationists in Austin Texas, explaining what steps could be taken to bring about a storm front. About a week later I saw headlines about heavy storms in Texas and adjoining states with Austin being among the hardest hit places.

These events just described really happened. However, they do not constitute proof that someone in Austin acted upon my suggestion and acquired a quantity of copper and placed it on a hillside facing the Gulf of Mexico, where the warm water generates copious quantities of water vapor. My opinion is someone did act on my suggestion and positioned far more copper than was really necessary to bring an end to the drought.

Or, being so close to the Gulf of Mexico is one place where only three or four hundred pounds of copper would be sufficient. The very thing humanity should by now be able to avoid appears to have occurred. This was the first letter I wrote to anyone about this possibility, and of course there was no reply. The possibility that flooding might happen had never occurred to me, and the whole news coverage of it left me stunned.

The Society of Separationists is or was an organization devoted to the separation of Church and State, started by Madelline Murray O'Hare, who fought some famous court trials concerning her children. Ms. O'Hare was an Atheist,and her children were required to attend public schools where prayers were read on occasion. She won the cases, and public school prayer in that school district was banned. That was in the 1950's. Madelline and two of her children disappeared and were presumed murdered around 1991, and around a half million dollars worth of gold coins belonging to the organization went missing.

I had decided to write to them in 1980 because they would not likely see any reason to keep this from the public, it does rather strengthen the position of atheism, and they issued a magazine. Texas seemed to see flooding on a number of religious holidays in the '80s. It was certainly not my hope that this should remain a discovery known only to a few, or for that matter, something that was used for less than ideal purposes.

I know this is all anecdotal evidence but these are anomalous events, these coincidences. Scientific jurisprudence requires the investigation of any anomalous phenomena, does it not? In 1982 I learned of the terrible drought ongoing in Somalia and Ethiopia, and mailed a letter to Saudi Arabian Bechtel Engineering, a division of Bechtel Engineering, a major U.S. defense contractor.

Naturally I never got the daily weather in Saudi Arabia or surrounding area, but in 2003 or 2004 an article appeared in New Scientist magazine online stating that the Northern Sahara had been shrinking for the past twenty years or so, from Mauritania in the west all the way to Eritrea in the east, and all countries in between.

Did someone acquire a quantity of copper in Saudi Arabia and place it in a high location and leave it there for twenty years? Could a multi-billion dollar engineering firm afford to and be capable of doing this? I guess the drought in Somalia and Ethiopia continued a while longer, which are a little south of Saudi Arabia, not west of there were precipitation appears to have increased for twenty years, so it was a miss when it came to that. About ten days or two weeks after I had written to Saudi Arabia, while watching the Weather Channel, the current weather in the Southern Atlantic was shown where a hurricane was tracking almost due east across the Atlantic Ocean straight for the Atlas Mountains in Western Africa, something that rarely happens. Usually hurricanes seem to track the other way, toward the Americas, in opposition to the prevailing westerlies.

That winter, 1982-1983, or the following winter,1983-1984 I forget which, I learned of a drought ongoing in Sydney, Australia, and wrote a letter to the American Embassy in Sydney with a brief description of what they might do in the circumstances they were finding themselves. Summer in Australia is during winter her in the U.S. Two weeks later I saw a small article in a newspaper about a storm that had hit Sydney, where it had rained "cats and dogs" for 24 hours straight. Once again, it looked to me as though someone had acted upon my suggestion.

In the summer of 1988 a drought was ongoing in the Midwestern United States. Barges were running aground on the Mississippi river, soybeans and corn were shriveling in the heat. I wrote a letter to "Successful Farming" magazine detailing the remedy for drought, and a week or ten days later intense storms hit the Midwest with Des Moines, Iowa being among those hit hardest by flooding, that city being the location of the headquarters of that magazine.

Did someone perhaps endeavor to see if what I suggested really worked? Some months went by after the letter to Successful Farming during which I went to the library to read the latest edition of the magazine to see if anything was written concerning this idea that had been forwarded to them and the recent weather there, and eventually gave up.

I went through this experience one more time, in 1992. I moved to Las Vegas, Nevada in August of 1991. On December 30,1991, after seeing something on television about the southwestern United States being in the grips of a seven year drought, I wrote and mailed a letter to Rolling Stone magazine on Wilshire Boulevard in Los Angeles, California. Therein I succinctly described how to implement a quantity of copper in a high location. January 6 or 7 saw rain across a wide area of the southwest. I was relieved when no flooding occurred to speak of, and began visiting the library to look through Rolling Stone magazine, expecting they would be progressive enough to handle the story. After several months of no article about anything remotely connected to weather modification, I stopped looking.

Five letters where it looked like someone may have acted upon my suggestion none of whom replied in any way. I e-mailed some people who were involved with trying to reduce smog in Mexico City and were using some kind of tall poles and electric current to induce the particulates to condense out,the subject of another article in New Scientist I happened to look at when I saw the one about the changes in the Sahara, around 2003 or 2004, and I do believe Mexico City has been rainy pretty much ever since, but I really don't know if that is a major change for Mexico City or not, and certainly no one replied to me.

Plausible deniability is the operative phrase, it applies to just about everyone. All weather from 1900 on is suspect. Tornado alley may exist in the U.S. for no other reason than some farming family in western Nebraska knew about this process for some generations. The current owners great grandfather knew Tesla personally, their farm gave them a unique perspective on the storms in 1899, who knows? Under the radar for over a century, the family has been up to mischief with the weather every spring and summer.

The entire human fascination with tornados has gone too far. The end result is always some disaster or another, yet the danger, and the filming of these events is now a national pastime. Almost as though the entire civilization would rather see this type of thing happening!

There are many indications that something unaccounted for has happened now, if we count Tesla and Hatfield, and throw all the strange weather in that has happened since 1900, all blamed on global warming, but possibly caused by curiosity killing the cat. The United States had a record number of hurricanes in 2005, including a good number that made landfall, and the following 2 years saw almost no hurricanes in the vicinity of the U.S.

Does Mother Nature change her mind that abruptly, or did someone place lead somewhere along the southeast coast to ward off hurricanes in 2006 and 2007, something that obviously wasn't done in 2005? There was a drought in the Southeastern United States in 2007 with water levels falling dangerously low. Could that drought be the outcome of lead placed along the coast to prevent hurricanes from approaching?

I eventually conducted further experiments on my own, in the desert southwest. From May of 2002 until probably early 2008 there were half a dozen objects resembling circular bird cages about 6 feet high each and 2 feet in diameter,all made of nearly pure copper, sitting on a hillside in the desert southwest of the U.S. The first two years there were only five such copper arrangements. The total weight of this row of copper tubing was between 300 and 350 pounds. The height of the hill upon which it was resting, on the southwestern side of this hill, is only 1500 feet or so above the valley stretching out to the south and west, so it had a small unobstructed space before mountains 6 or 8 miles distant block the signal from the copper.

The quantity of copper needed to be greater, and the elevation of its location needed to be higher, to see more robust precipitation. Nevertheless there were observable effects, including numerous small

Doesn't show the band of clouds overhead

thunderclouds dropping rain in small areas surrounding the hill with the copper. Of the 10 or 12 times I have driven past the location within a few miles, small rain clouds covering just a few hundred yards have splattered my windshield about 75 % of the time.

It was not unusual to be hit by three or four of these small storms 10 miles either side of the location.There was quite often a persistent narrow band of clouds directly over this location, with the trail of clouds only a few hundred yards or so wide stretching 3 or 4 miles in either direction from the hillside where the copper rested, and in line with how it was positioned.

When I first hauled the materials up the hillside to where they rested, I laid it all on a flat ledge horizontally and didn't return for two weeks, at which time the wind in the area was excessive. Returning two weekends later, carrying everything further up the hillside and positioning them vertically happened next. Once the circular coils were tied together with copper wire to straight bars of copper tubing, and standing vertically, cloud development became much more frequent in the area, and the winds settled down.

On Labor Day around 2004,I and two partners in mischief ascended to the copper with more circular copper tubing and assembled the sixth bird cage type object. We borrowed from the abundance of straight bars of copper tubing that was attached to the original five contraptions. The three of us carried all six up the hill a little further, and began our descent at 4P.M. that Monday afternoon. Rain began at 4P.M. that Thursday over a wide area including where I live, and since there is an hour time difference between where I live and where the copper was, the elapsed time was closer to 73 hours.

Once the first edition was pretty well along I made a journey to the sight where the half dozen copper devices were left, with a friend, who also happens to own the property, with the intention of taking some pictures to add to this book. After climbing around the entire hill, we found where the site had been,but found only a hack saw, and one of the copper wires that we had used to tie one of the copper constructions to a nearby rock to keep it from being blown over by the wind, with the loop that had gone around the rock still tied.

So the experiment came to an inglorious end when someone noticed the strange looking objects while hiking through the area, looked closer, decided that it was something valuable, and stole it. It must have come as a shock to whoever found those things, to see a half dozen odd looking contraptions in the middle of nothing but rocks, sand, cacti, and sagebrush. I do have a few pictures of the experiments though, included at the end of the book. I was not aware of the existence of these photos when the first edition of this book was made available to the public. My friend, the land owner, had an old cell phone where the pictures in the back of the book came from. Those pictures were taken by him, he is not in any of the pictures. My other friend with the sunglasses is, and yours truly. One of the pictures shows the site when we first came upon the copper, and it shows how at least one of the objects had been blown over by the wind.

The picture looking down the hill over my friend's shoulder shows the trailer my friend the landowner lives in, and several vehicles. It takes about an hour to walk down from there to his trailer, and a little longer going up. The theft we think took place in early to mid 2008. The last visit, when we were unsuccessful in finding the copper, was in September of 2008. If one were to look at the weather in Kingman AZ from 2002 through 2007, it should show above average rainfall. The location was about ten miles toward Las Vegas from Kingman. Come over the Hoover Dam on the 93 southbound, hit the double lane section 12 miles distant, and the weather was different back then, clear to the other side of Kingman.

It is apparent now at the end of this chapter that there could be a great many other ways of positioning copper that I haven't tried. Solid bars of copper, multiple placements of copper of various kinds over a few hundred square mile area, these are intriguing options to explore. Could very well be that a half dozen two or three hundred pound copper placements spread around a geographical area would cause just the right kind of cloud development to prevent tornadoes and huge thunderstorms and hail from occurring. The clouds wouldn't clump into cumulo-nimbus clouds, but might rather be more general cloudiness culminating in precipitation.

Chapter 9: From The Beginning

Question Everything. Hot Potatoes First
JB

My interest in copper and the atmosphere began with a book titled "Modern Physics and Antiphysics"[10], which I read back in the 70's. I became fascinated by the part where the author discusses "The Absolute Frame Of Reference", and the search for it. What we can distinguish about things and their movement in space always comes from a frame of reference that is relative to another observer, or relative to some stationary object. Though that stationary object is moving, it is moving right along with an observer, and is thus apparently motionless.

Indeed, anyone anywhere can testify to the fact that, though we are moving at incredible velocity, we could be sitting at a restaurant enjoying dinner, and everything around us appears motionless. The other diners appear to be enjoying themselves, the wait staff has no difficulty moving from place to place, the famous picture on the wall observed sitting down at the table is still there three quarters of an hour later, with dessert and coffee. Nothing has moved, but that is only relative to an observer in a fixed location. Everything moved, but it all went the same direction. The frictionless passage of the Earth through space goes undetected, the actual velocity and direction unknown, and the speed quite considerable.

Our frame of reference is our own Solar System, generally, since all within it is in orbit around the Sun, and following the Sun as it orbits the Milky Way. Knowing the absolute frame of reference would enable one to distinguish between one square foot of space and it's adjacent square foot of empty space. Areas of space are things that are not labeled in any way, are completely identical, and consist of mere emptiness.

If one could account for all motions that the Earth is undergoing, and travel backwards at the exact speed and direction,you will have reached a point of absolute motionlessness, with the exception of your bodily functions, but who knows? Maybe something weird would happen then, when you are no longer moving in space. There has been discussion about zero point energy that ties in with the absolute frame of reference and knowing one's exact speed and direction as well. Anyway, the discussion began with Michelson and Morley, and the search for the "ether wind". I think that was the title of the chapter.

Michelson and Morley were searching for an ether wind, a kind of stratum in empty space through which things pass, theorized to exist to explain gravity, or the passage of light through vast reaches of space. Einstein's discussions of how one observer, being in a different place from another observer, may see an event in an entirely different frame of reference as actually occurring differently from the account of the other observer, was also mentioned.

Anyway, there were various hypotheses about the subject matter, of the void, and was anything in it.Also how, indeed, does one distinguish between one square foot of empty space and another, since we could be sliding any which way between any astounding number of empty square feet, and that thus far we have absolutely no way to distinguish one from another. The Solar System, in orbit around the Milky Way, is moving somewhere around 500 thousand miles an hour. One could travel to the moon and back in an hour at that speed. The problem defined, as time went on an idea came to me about how one might possibly conduct an experiment to find the absolute frame of reference. The problem remains unsolved to this day, and many think it insoluble. Obviously I didn't solve it either.

First of all, if one were to try such an experiment, the experiment would take place in a fixed location on the surface of the Earth at some specific time. Each day, wherever we are going makes a full revolution from our perspective on the surface of the Earth. Though the Earth continues in one basic direction, at noon it is 180 degrees different from whatever direction and speed it is going at midnight, to an observer in a fixed location on the Earth. For example, suppose at noon the direction of the Earth is straight up over our heads. At midnight it would be straight down beneath our feet. This is the most rapid change. Every four minutes the Earth has spun one degree on its axis.

Starting with the Big Bang, all of the things that eventually become our Solar System are thrown out at enormous speeds along with everything else. We may still retain some inertia form that event, so that is one motion, if we assume the collision of the two super massive black holes that created our known universe,or whatever theory about it coming out of a singularity, was absolutely motionless at the time.

The Big Bang could have been sliding one direction or another while it happened. One of the black holes that collided with the other could have been larger than the other. The impetus of the larger black hole over the smaller would be impossible to calculate. Thus, determining the ultimate direction of things might ultimately prove impossible mathematically,since one could never have the figure for any other possible motions. After the Big Bang, there is galaxy movement within the cluster of galaxies that includes the Milky Way, our rotation as a star system with planets around our galaxy,our rotation around the sun, and the Earth's rotation on it's axis.The pull of distant objects another.

Given that there are 5 or more different motions to calculate, and high speed velocities, the curving path would be as near a straight line as one could get for any short distance. That it is not a perfectly straight line would not reveal itself for thousands of miles. That consideration understood, one could conceivably find where exactly the matter in our solar system is going by accident. Lined up in the nearly straight line that all the local matter is traveling in, there could be static electricity left in our wake, and that could eventually be detected and collected.

This idea tied in with the discovery withheld from the world in the novel Atlas Shrugged. The thing that intrigued me about that book was the invention. The invention consisted of some device that captured ambient static electricity and converted it into electricity. The ambient static electricity would be left in our wake, was my conjecture, as we sped through the universe at incredible velocity, perhaps.

I even hypothesized at the time that since space is currently indistinguishable, in so far as telling which identical area of space is which, and that the Big Bang being such a colossal explosion, perhaps the matter flying apart exceeded the speed of light, and could even be doing so now, since the universe seems to be expanding very quickly.

Everything is going that way right now

For all we know our limited viewpoint won't tell us exactly how fast we are going anymore than where or in what direction. What if one knew the absolute frame of reference and had just the right type of arrangement which would be easy to make with a length of copper refrigerator tubing, aligned so that as it flew through space exactly so, it was in line with our actual direction through space and would be in position to capture the static electricity? Maybe that was how it could work.

I thought that would be the only way to get an inkling of the Absolute Frame of Reference, by deliberately stumbling upon it by giving enough different tries to a length of copper, at different positions and angles. What was the rush, anyway, I probably would never find it, but even if I didn't I would still have the copper, and could continue with it,maybe refine the experiment with knowledge of the direction of the solar system through the galaxy at 12 noon in my neck of the woods, get a voltmeter, etc. That was how it all began.

Having determined it would probably take years, and pretty much sure that I would need to eventually refine the experiment to accomplish anything, but impulsive enough to try anyway,I acquired two continuous lengths of circular copper tubing that could be stretched out to a length of twenty feet or so, each loop of coil about a foot apart, and a foot and a half in diameter. I think the coils were three eighths inch diameter hollow tubing, 99.997% copper. My main conjecture was that ambient static electricity might accumulate in some circular tubing just because copper conducts electricity so well, something that could possibly be captured better along the length of copper depending on how it is positioned, so I stretched these two out the length of an upstairs bedroom, and at the lower end was a car battery out of power, and the upper ends were by the one window which I opened to the north, and the one window I opened to the west.

I think I noted the time to myself, and three days go by, and there was a storm moving in to the neighborhood. Being fascinated by the weather, I went out in the back yard, and was watching the lightning, and listening to the thunder. The storm was coming in close, the thunder was getting louder,and heard in less time after the flash, when suddenly the loudest clap of thunder I have ever heard shook me to my roots, and in that instant, the translation of the name Ayn Rand into the phonetic pronunciation "Aye 'n Rained" like an Irish person talking struck me, and I finally perceived what had eluded me up to that point. My mind raced with the possibilities. There was the start of my realization that weather modification might be a possibility, and had probably long been known, and withheld from the public.

Now it seems to me it was a dumb experiment that lasted just three days with no chance of succeeding unless some sensitive measuring instruments were used but I did it, and that blast of thunder sent me away from ambient static electricity and the absolute frame of reference and on to weather modification and collecting electricity from lightning in an instant.

There I was unaware of the possibility that the copper tubing by those two open windows might have had something to do with the fact that this storm was bearing right down on us from the northwest, I had forgotten all about them, when that clap of thunder and the thunder bolt of realizations in my head happened at the same instant. It was plain that the copper must have been causing a path of least resistance which the incredibly numerous atmospheric particles, having electrons, would follow.

Perhaps it may have had some short term impact on the storm in the area, maybe enough to bring one lightning strike quite near, but it was certainly too small an amount to actually generate the storm, at least that is what I thought.

The actual beginning of all this began with an idea about something else entirely, then, and the first events where I experienced experiments with copper involving the atmosphere were quite inconclusive. The epiphany, though, the thunderbolt, had me convinced I was on the right track, that something conductive, large, in a high place would create a path of least resistance for atmospheric components to more easily slide along. Eventually I learned of Tesla's exploits in Colorado Springs in 1899, and the three decades of wet weather in the middle of the United States that followed.

At this point, I would also have to add that my brain provided me with a sudden insight during that blast of thunder; I tend to trust my inspirations if and when they happen. To have done otherwise would have been foolish in the extreme. Whatever electrical events took place in my mind to cause me to be conscious of what I became conscious of were going on at a deeper level of consciousness where much more information is being cross referenced. The sum totality of my being told me the truth, and I believed it.

I began to think of correlations in the real world that roughly corresponded to what I thought was taking place. A car traveling on an icy road, for example. The car reaches a curve, the driver turns the wheel, but there is a patch of ice on the road. The car continues along the path of least resistance, there is not enough friction for the front tires to grip, turning the car, so the car goes off the road. Another one involved just the copper reducing the amount of static electricity in the air, and that gave rise to the more conducive path, less electrical resistance. That area closest the copper,with little or no static electricity at all, creates a small vortex that snowballs, grow and intensifies, into a low pressure system.

Mostly what went through my mind was a rule of thumb in physics, basically why water doesn't run uphill. If energy is required for something to not follow a path of least resistance, in short, for water to run uphill, it won't do it. It will take the path requiring the least energy. The conclusion one could reach is that just about any small atmospheric component wouldn't expend energy to avoid the path of least resistance, and are quite likely to gravitate towards the path of least resistance, that being the easiest path to follow, and also one that is being followed by numerous charged atmospheric components.

All of this was conjecture long before I ever realized that atmospheric nitrogen and oxygen, 99% of the known air, were electrically neutral entities unswayed by the electromagnetic force of the copper themselves. Prior to that, I was assuming that the extremely small size of the free floating gases in the atmosphere, and the fact that they do contain electrons,sent them along the path requiring the least energy.

Once I learned that the nitrogen and oxygen were unresponsive to the electromagnetic force of the copper, it took me a while to see that dark matter and dark energy could tie in with what I perceived to be happening in the atmosphere. That didn't happen until some time in late 2007, a time when dark matter and dark energy had been around on science shows for a while, not the case in 1980. At that point the first edition of this book was already underway.

The next significant find for me was the peculiar nature of the water molecule itself, now in my estimation the biggest contributor of all to what is contended to be taking place. That came at a time when the current edition was nearing completion.I still haven't accorded that as much attention as it deserves. For one, relative humidity can be quite misleading. Water reaches the saturation point at 100% relative humidity, but the actual amount of water by weight in the atmosphere then seldom exceeds 3%. Water molecules are lighter by weight than nitrogen and oxygen isotopes, so there would need to be more of them to make up the percentage of air by weight that they occupy. Still, if the actual humidity by weight of water in the atmosphere at the start of an experiment is 0.25%, and it must reach 3% for rain to begin to fall, there really isn't that far to go, when one considers that the water molecules themselves are going to begin accumulating along the path of least resistance. Just water molecules flocking to a location may by itself be responsible for the observable effects, though falling barometric pressure would likely involve primordial specks to some extent as well.

To properly conduct a hot or miss type of experiment to find the absolute frame of reference such as I had begun to try to carry out and quickly abandoned back in 1980, one would need a huge room like an aircraft hangar or a large pole barn, a stand to place super conducting material on that one could turn in any direction with control mechanisms, a one hundred foot long super conducting composite material in a straight circular tube, sensing equipment all along the length of the super conducting material, and a computer to monitor the whole event, all the while calculating exact geometric position in the universe, motion of Earth's axis, orbit around the sun, and the orbit of our solar system around the galaxy.

The Earth turns full circle in 24 hours, so 360 degrees divided by 24 hours gives us 15 degrees of arc every hour. If we are traveling around the Milky Way around 500,000 miles per hour,that distance we travel in one hour divided by 15 gives one degree of curvature every 33,000 miles or so.Just from this guess, one mile would reveal 1/33000th of one degree of curvature and 100 feet is only one 52nd or so as long as a mile. That would truly be almost as straight a line as one could possibly get, without actually being straight, for the 100 foot length of super conducting material.

Taking the 500,000 miles an hour or so that we are traveling around the Milky Way, that computes to over 138 miles per second. What fraction of a second elapses while the Earth and everything on it slide one hundred feet? There are 733,300 feet in one second of travel at this speed. So that is 7,333 one hundred foot lengths in one second, so the time elapsed is 1/7333 of a second. That comes to 1.363 ten thousandths of a second in a straight line, changing slowly, one degree every four minutes, besides some other much more gradual changes. Zooming along we are, and other curves in the path would be very gradual in comparison to the 24 hour spin of the Earth on its axis. As we orbit the Sun, we slowly curve around it, but the entire journey is a year long to go a full 360 degrees and covers many miles,so the degree of curvature has to be pretty small. Just considering that 30 degrees of curvature takes a month, then one degree of curvature is achieved in about a day as we go around the Sun. One degree of arc takes 4 minutes as the Earth rotates on its axis.

How straight a line for 100 feet with one degree of curvature every four minutes? 4 minutes has 240 seconds. 240 seconds has 2,400,000 ten thousandths of a second. Two ten thousandths of a second would reveal 1/1,200,000 of one degree of arc. My guess is that is so close to a straight line it is indistinguishable from one, and the actual figure would be slightly lower than the estimate from two ten thousandths of a second, we had 1.363 ten thousandths of a second estimated elapsed time to travel 100 feet.

The orbit of the Solar System around the galaxy takes 100,000 years and almost 440 trillion miles to go full circle. 360 degrees of arc divided among 440 trillion miles won't be much, one degree of arc is easily more than a trillion miles long, much longer than 33,000 miles.

We could be going faster or slower once all combined movements are accounted for, but the likelihood is we are traveling even faster as a component of the Milky Way as it orbits the local cluster of galaxies. The Earth is traveling around the Sun slower than the Sun is traveling around the Milky Way, and the Sun must be traveling slower around the galaxy than the galaxy around the local cluster of galaxies that it is among.

The curving path of the orbits of things would also be more gradual the larger the thing or group of things is. We can't expect galaxies to turn sharp corners. The pull of more distant objects is hardest to estimate.

One would need to know to know the extent to which the Milky Way is being pulled toward the unknown attractive force in the Andrew Kashlinsky article. Some of the galaxies nearest this probable black hole are moving towards the thing at 3 million miles per hour, others farther away much slower. If the Milky Way is also headed there, that speed might be our fastest and primary direction.

Then one would program the robotic positioning to move position every two minutes, or some such small time, and let the thing run itself for several months, return, analyze the data, find no anomalies, run it another six months, and maybe find something odd occurred during one two minute stop, bring up all the sensor data recordings from the event, calculate the exact position, the time it occurred, the day of the year, etc. Realistically one could probably try the experiment on a miniaturized scale, with just a three foot long bar of super conducting material, on a work bench. In that instance, three feet of travel is all that would be involved, so without question a straight line for so short a distance.

I still get the distinct impression that the experiment might actually work. Acute sensors would pick up more static electricity when in alignment with the path we are traveling upon, since static electricity is not bound by gravity; the main problem is getting the two in alignment. Chances are the amount of static electricity that could be captured wouldn't be enough to make it economically feasible to produce electricity on a large scale. But, if there is a small quantity there, and the absolute direction is found one time, it could be tracked continually from then on.

The first positive reading gives experimenters a starting point, and losing track of the direction would thereafter be temporary, and eventually worked out over an extended period of time. Then, using the direction of things, which is then known, and numerous astronomical readings, a computer could triangulate the exact velocity of things in the Solar System, even stars in other galaxies. All one would need is a little time lapse to triangulate velocities. So the question of the absolute frame of reference gets solved, except that the finding of the direction, a starting point from which the velocity could be eventually calculated, is currently only theoretically possible as here stated and technically difficult.

Static electricity may not be bound by gravity, but there is every reason to think that it would, upon creation through friction, retain whatever inertia the things possessed that generated the static electricity. In other words, static electricity, upon creation, isn't going to just stop on a dime at that moment, it will continue along with everything else in the vicinity. There might not be a sudden spike of electricity along a length of super conducting material if it is lined up with the true direction of everything traveling in the vicinity. Then again, one could argue that as time goes on, some static electricity could still accumulate in our wake. Those bits of static electricity already in existence, cast adrift with some inertia, eventually could fall behind the rapid progression of the Earth and surrounding Solar System. Supposing the Solar System and all in it are being pulled toward the "Deep Drift" object at an increasing speed, the static electricity, once it is on it's own, won't be getting pulled there, and will slow down. The question of whether a greater amount of static electricity would be detectable along a straight length of super conducting material at only one particular direction of alignment isn't easy to answer.

I am no astronomer, but the "Deep Drift" object was said to be in Saggitarius, some constellation in the sky. Where that is in relation to where an experiment is conducted might help narrow down the range somewhat. Calculate what direction the Milky Way orbits the local cluster of galaxies, add where the Solar System is going as it orbits the Milky Way, and add that angle to the drift toward Saggitarius, all the while factoring in Earth's spin on it's axis. Of all possible different directions in space, perhaps 80% of them could be eliminated from the search. The remaining 20% could have a pleasant surprise somewhere.

It would be quite an accomplishment if the scientific community could solve the problem of the absolute frame of reference. Much more would be within our grasp were all scientific disciplines attuned to exactly which way and how fast we are traveling through the universe. It could have implications in chemistry, new methods that would make it possible to make certain compounds more easily. Medical compounds once difficult to make and expensive become affordable to everyone. Who knows what one could do with buckminsterfullerenes if one knew the absolute frame of reference. Manufacturers of micro processors could make use of the absolute frame of reference to more finely hone the internal components of computers. Any type of nano material or process involves very tiny things and events, and the absolute frame of reference could factor into how things could be more easily assembled on the molecular scale. The medical field may be able to integrate the new information when creating replacement organs, making more finely specialized parts.

Soon after the blast of thunder in June,1980 when I first realized these new possibilities I soon brought those copper coils outside to try to see what, if any, effect they might have, and these experiments were dumb too, because I think I had only 40 pounds of copper, and no real high place to put them. Even so, I think they may have contributed to cloud development, and when I placed them outside cloud development did occur, and seem to peak in intensity around 72 hours after putting them somewhere, and usually look as though rain could come, but usually little or none. I did add some to it, over the early 80's, to where I was carrying around about 75 pounds of copper. I usually only ventured out when forecasts called for cloudless skies.

All I can conclude is that nothing I did at that time actually proved anything definitively, but eventually I came to realize a path of least resistance really does seem to be at the heart of the matter,one involving the water molecule and the primordial specks, and not N_2 and O_2, although the nitrogen and oxygen appear to be involved via gravity. So I realized it would need quite a bit more copper than what I was experimenting with back then in the early 80's to actually produce a decent storm system.

I had seen what had happened in Texas, Australia, and Iowa where drought ended suddenly shortly after my letters were sent there. After a while I concluded those people had probably tried a ton of copper on some hillside, maybe even two. Copper was around $1.50 a pound back then. Two tons of copper cost $6000 back then, petty cash for a lot of wealthy corporations and people. I never advised anyone to try any more than 500 pounds, at least on the first try, and all those events got me eventually contemplating further what had probably happened in those places.

Whenever an amount of copper in a high location exceeds a half ton or so, hydrogen nucleo-synthesis in the atmosphere due to the intense concentrations of primordial specks probably escalates to the point where hydrogen pairing and meeting single oxygen isotopes produces incredible volumes of water in the 72 hour period in which the storm front is developing. When the precipitation arrives it is too much. The more copper there is, the greater the intensity of the electro-magnetic waves that propagate outward from them. Water molecules will begin gathering in ever increasing numbers. The atmospheric components would be pulled that much more vigorously along by an even greater concentration of water molecules and primordial specks, and the range that the effects reach would grow larger as more copper is used. All the theorized effects would increase or decrease in correspondence with how much copper is used.

It is also worth mentioning that when an experiment with copper on a hillside begins to bring about increased quantities of water molecules and primordial specks, the H_2 in the water molecules is probably still capable of spawning further H even if now connected to an oxygen atom. Such an experiment would have brought together all the necessary ingredients for increased hydrogen production. Any increased hydrogen production occurring in the atmosphere of the Earth would invariably lead to more water molecules, as was observed previously, the rate dependent on how many new hydrogen escape into outer space without meeting up with an oxygen atom.

Maybe there aren't any accurate predictions to be made when one uses quantities of copper beyond the 350 pounds at most that was used in my experiment. I never experimented in enough places or with enough copper to be making definitive statements, or so the argument goes. Perhaps there are some unexpected complications in some locations on Earth. Nevertheless, something, not nothing, is definitely happening when copper in the hundreds of pounds is placed in a high location. The only way to find out, for example, if the tornados in the middle of the country could be reduced or eliminated entirely would be to conduct a few different experiments. Denying that copper so placed does anything merely delays the inevitable, which is the truth coming out, and getting put to appropriate use.

Chances are a certain location where 650 pounds of copper strategically placed caused just the right sized storm to pass through, upon subsequent experiments with the same amounts of copper, different results were obtained. The first try was during a weather system quite different from the second one. Several hundred miles to the west, a very large low pressure system was already in place on the second try, and with the added drop in barometric pressure from the copper of the second experiment, the storm system grew out of control. That type of thing can happen. Experimenters need to be cognizant of current weather systems and factor in low pressure systems inbound. In this particular instance, the experimenters should have reduced the amount of copper by as much as a third to minimize the chance of disaster. Just how predictable these types of experiments can be will eventually be found out as more experiments are conducted.

Moving forward with something that offers possible solutions to a plethora of problems seems the best choice for humanity. If water boiled at a lower temperature would we refuse to use it? Just because atmospheric experiments involving copper have a point beyond which dangerous weather could occur and that point is pretty easily reached, doesn't mean mankind couldn't live with it. We are already living with it; to continue to deny that dangerous weather could be man made, or might be avoided, is fatalistic. The idea is to promote the advantages of having peaceful weather with abundant precipitation worldwide. Going with the idea puts one in the position of being able to do something about unwanted weather.

With the certainty that the water molecule itself is responsible for most of the observed effects, and the theory that the path of least resistance pulls primordial specks along the path of least resistance, some experiments might clarify what is indeed causing changes to the atmosphere. If huge electromagnets work even better than copper tubing at causing clouds to form and rain to develop,then it would be the water molecule most responsible for doing that; if such experiments were less successful, the copper in a row would be more of a factor than magnetic attraction. Should both those methods give inconclusive results as to which one was more efficient at herding water to a specific location, then it is a little of both. We may be able to narrow down the possible causes a little.

Similarly, one could compare the effectiveness of placing lead sheets in a high location spread north to south with placing on high a large electromagnet with a negative charge to repel water molecules. Earlier the pie got divided up with 40% to the water molecule, 30% to the primordial specks, 15% to the Jet Stream, 10% to the static electricity, and 5% to the Earth's magnetic field. Based on what is more effective those hypothetical slices could change considerably. 70% of the effects are due to the path of least resistance along which both water molecules and primordial specks would travel most readily, currently.

If electromagnets with different polarities prove even more effective at attracting water and repelling water and successful predictions could be made about what was likely to happen in situations like this, then the path of least resistance wouldn't figure as heavily into how precipitation happened, or was prevented from happening, but would be more due to the negatively charged oxygen nucleus in the water molecule, the larger component, being attracted to positive electrical charges and being repelled by negative ones.Conversely, should copper and lead prove to be the more effective weather modifier, one would have to stay around the original estimate of 40% to the water molecule, 30% to the primordial specks, etc. My guess is copper in rows of tubing pointing skyward generally, and lead in thin sheets high up from north to south would yield more predictable results. A strong electromagnet might bring too much water too soon and conversely send water molecules away so strenuously as to cause weather problems elsewhere. Beside that, using copper and lead would be cheaper and not involve expending electricity, but rather includes the possibility of its capture.

Perhaps the early experimenters in the first three decades were using electromagnets and these were responsible for all the flooding, and the dust bowl was caused by a decision to reverse the polarity of the electromagnets. Since a lot of things seemed to go awry back then the copper and lead seem a bit safer since what they are likely to do are predictable, provided an experimenter in a certain area is the sole experimenter, and caution is observed.

Needless to say, anecdotal evidence does not constitute proof. However, when there is a preponderance of events where an experiment appeared to have been attempted, and each of those events saw a change from a drought situation to a condition where water became all too abundant quickly, one should, at least, begin to wonder. The properties of the water molecule being such as has been shown in this edition being more likely at the bottom of it all, or at least a significant factor, it makes me wonder how someone who has read all that has been presented here could conclude otherwise than the author. True enough that the first edition didn't even mention anything about the features of the water molecule. Added to everything it puts the idea more in the inevitable category than in the possible category.

For thirty one years I have investigated what intrigued me in the summer of 1980 concerning a path of least resistance being created by a quantity of copper in a high location. The conclusion reached is precipitation is easily obtained virtually anywhere on Earth, during any season. The process is as nearly tested, tried, and true as any cautious investigator with limited resources could determine. Extrapolating that lead would have the opposite effect is not unreasonable. What happens from this point on in the entire world should be an improvement. Steering clear of weather disasters will eventually be the norm.

Chapter 10: Suppositions and Predictions

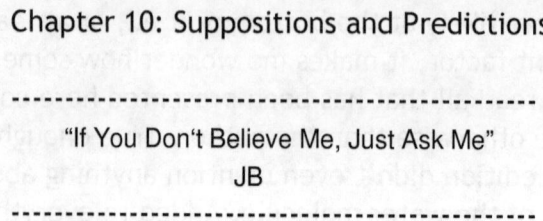

"If You Don't Believe Me, Just Ask Me"
JB

Last but not least, a firm record of the entire Primordial Speck Theory needs to be written up, with all its suppositions and predictions. That way, we can all see in the fullness of time whether the predictions are fulfilled, and if the suppositions it makes hold true. Someday perhaps more sensitive detecting equipment will be designed that will distinguish three or four different particles in random space that previous to this time were below the threshold of detection.

If that happens, a one particle hydrogen precursor theory would obviously be wrong, and not stand the test of time. Most of the suppositions made by the theory would all probably still hold even if the number of extremely small things was more than one type. The different types would combine in whatever way to become hydrogen whatever the dark matter and dark energy are, and we are currently in a position where only theory will shed any light. In favor of simplicity the theory goes with one fundamental particle, since the most pulverized thing possible doesn't seem likely to have choices.

Supposition 1: That undetected dark matter and dark energy in the known universe are primordial specks, precursors to hydrogen made of combinations of matter and energy too small to detect, probably one fundamental particle in astounding abundance.

Supposition 2: That these primordial specks are most likely magnetic monopoles. When primordial specks are in sufficient concentrations they transform into hydrogen through the stable hydrogen isotope, H_2, serving as a template. The likelihood is that we will never know how small each individual primordial speck is, nor the steps it takes to unite with others of its kind and become a hydrogen atom, comprised of a proton with three quarks within, and an electron in orbit around it. Information is somehow incorporated into very,very small things in the space between two identical hydrogen atoms that assume the shape of a larger thing, becoming an unity.

Supposition 3: Being magnetic monopoles, the primordial specks are unstable and energetic. They will accumulate along a path of least resistance such as can be created by placing a quarter ton of copper as earlier specified near a mountain top facing the prevailing westerlies.

Supposition 4: The accumulation of primordial specks in increased concentrations along a path of least resistance and the resultant nucleo-synthesis of hydrogen in the atmosphere will result in newly formed hydrogen atoms pairing up as stable H2 isotopes, rising to the ozone layer, joining with a free oxygen atom there, and in great quantities becoming water molecules that join developing clouds.

Supposition 5: The primordial specks gathering in increasing numbers along the path of least resistance, all traveling in the same general direction, along with increasing numbers of water molecules, will gravitationally pull the electrically inert N_2 and O_2 that accounts for 99% of the known air along with all of them along the path of least resistance. The other 1% of the known air, argon, methane, neon, carbon dioxide, etc., will also succumb to the gravitational pull of the mass of the water molecules and primordial specks, and join the schooling cascade. All this, of course, results in falling barometric pressure, cloud accumulation and ultimately, precipitation, usually around 72 hours into the experiment. Supposition 4 also figures into the clouds and precipitation.

Supposition 6: The combined impact of suppositions 4 and 5 along with perhaps some other additional factors helping such as the Jet Stream, static electricity, magnetic field changes, etc., proves that a path of least resistance can be created with a row of copper tubing in a high location, resulting, if the experiment is properly conducted and uncontaminated by other nearby experiments, in precipitation occurring 72 hours, give or take 8 hours, into the experiment. Also that cloud development will start promptly within an hour or two of a sudden insertion of copper of sufficient size near a mountain top, and build in intensity until the release occurs in the form of precipitation. The polar nature of the water molecule may prove to be the most decisive factor in producing the effects described, but probably isn't enough by itself.

Supposition 7: That since we have established that dark energy and dark matter is in the form of particulates that are a combination of both matter and energy, the excess of the dark energy over dark matter can be explained by the 52% excess dark energy consisting of the gravitational pull of massive objects further away from us than the known universe. This gives us a 90 to 92 percent percentage of everything in the known universe being primordial specks and 8 to 10 percent being known matter, hydrogen on up the table of elements, along with a few stand alone particles like neutrinos, and released energy from matter in various forms.

The increased acceleration of the known universe is explained by this and it is predicted that other objects of mass will be discovered outside the known universe. This may have already happened, though, over the last few years with recent astronomical discoveries that are still being investigated, so my thinking is that the two recent science articles about things possibly outside the known universe will eventually be confirmed. Right now the idea that 52% of the dark energy is in the form of distant objects converts the dark energy into things, then, that are a combination of matter and energy, just like here, it is simply that we are getting no information from them other than their gravitational pull.The mass of the hypothesized more distant objects have never figured in to calculations because we have no way currently of finding it out. We only have the calculated strength of a force trying to pull the known universe apart. This supposition, in this second edition, doesn't predict objects outside the known universe will be discovered in the next 50 years, as the first edition did, since the first edition prediction may have already happened before the book was written.

Supposition 8: That space has not, as is thought by many, been shown to admit of alteration in any way.

That photons, a form of energy with no mass, should be bent by massive objects in their vicinity, proves nothing other than that the photons changed course, not that the space changed. A more thorough theory of gravity with space as a fabric thrown out will emerge from this within 20 years. Space will be concluded to be absolute, eternal, and it would be impossible to remove nothingness from somewhere.

Supposition 9: Particles do not pop out of zero point energy or some such things, they coalesce into hydrogen from smaller particles in all probability within a stable hydrogen isotope. These primordial specks will eventually be concluded to exist, and be regarded as the most abundant thing in the universe by almost all scientists and literate people in 40 years.

Supposition 10: Given the considerations stated in Supposition 7, the known universe is not the entire universe. Given that, our known universe need not collapse back in on itself, and accelerated expansion of the universe will probably continue, and eventually the cold, lifeless things that were our solar system will be absorbed by a super massive black hole, which will eventually collide with another, spawning another 100 billion or so galaxies, so the entire universe will keep existing, and we will never know its full extent.

Supposition 11: That an experiment with copper in a high location, yielding increased concentrations of primordial specks would show more hydrogen atoms in random air samples taken 72 into the experiment than an experiment of the same type that instead used rows of lead sheets positioned from north to south facing the prevailing westerlies,with random air samples taken 72 hours into the lead type experiment of identical size as the previous experiment.

For successful tests of that prediction, however, one must realize that in the copper part of the experiment, the barometric pressure will in all likelihood be much lower, the relative humidity much higher, and it will probably be raining. Thus, some of the hydrogen found at that moment might just have come from a water molecule breaking apart, so one would have to factor in an estimated quantity of those. I really don't know if water molecules break apart easily or not. One would also wonder how many water molecules have brand new H_2 isotopes, or half new, half old, but determining that is not possible.

It is predicted that numerous experiments of the type mentioned will be carried out, and confirmation that additional hydrogen is present in greater parts per billion in the copper part of the experiment beyond an amount correctly calculated for water molecules breaking apart will happen within 30 years.

Suppositions 12: Within thirty years, an experiment with copper in a high location with a vacuum chamber nearby will discover additional hydrogen in the vacuum chamber after it had been placed, with some hydrogen atoms within to serve as templates, at the site of the experiment with copper. A similar experiment with a vacuum chamber that had been emptied as completely as possible of the same size and left at the same site will not be found to contain any additional hydrogen. This will prove without a doubt that new hydrogen requires already existing hydrogen to serve as a template, and also proves that hydrogen production in the early universe began slowly, and is now still ongoing, mostly within stars.

Supposition 13: Some time in the last three decades of the twenty first century,a mathematician will calculate accurately the smallest magnetic monopole the universe can possibly make, and computer simulations with the new data inputted for the mass of the primordial specks will show a high degree of accuracy in predicting actual conditions. With the new information as to the size of the magnetic monopoles and the requirement of H_2 templates for hydrogen nucleo-synthesis to occur, the rate of current hydrogen production will be calculated and found to be still quite actively taking place.

Supposition14: That the dangerous storms that brew out over the oceans and are known by the names hurricane, typhoon, and cyclone could be wholly eliminated if mankind cooperated with each other and diverted a substantial portion of the water vapor and static electricity out over the oceans towards land. Additional drainage of the oceans could be brought about by positioning more than the usual amount of copper near the north and south poles. This last mentioned activity would also rebuild the polar ice caps. The prediction is a sharp falling off of the aforementioned storms in the next two decades.

Supposition 15: That tornados could be eliminated to somewhere within a 98% reduction, year to year, with a little experimentation with different ways of creating a path of least resistance. It is predicted that the prevalence of tornados will fall off sharply within two decades. Flooding and drought also become much less troublesome in the next two decades. It is predicted that a barometric pressure hotline will come into being some time in the next two decades. It is also predicted that this hotline will no longer exist in 50 years, along with the weather channel.

Supposition 16: That the insurance industry and the various branches of government in the United States will revise the term "Acts of God" to "Acts of Nature With Possible Human Involvement", or some other more accurate term when referring to weather events by the end of the 21st century.

Supposition 17: That since we have restored space to her former glory, and space/time being a mistaken idea, or something applicable to equations but not reality, the Big Bang didn't occur as theorists now maintain it did with space/time collapsing into a point together with all the mass of the entire universe contained within, but was the collision of two super massive black holes, in a very empty quadrant of the much larger universe, since these two super massive black holes that collided vacuumed up all the matter in the area to a distance of 10 billion light years in every direction before colliding, though a trail of particles most likely followed both black holes as each accelerated toward each other, including stable hydrogen pairs.It is predicted that Einstein's theories will still hold, with the new condition that time may be variable, but space is not.

Supposition 18: That computer simulations with the suppositions listed above will likely yield an accurate picture of reality, and thus be able to successfully predict atmospheric conditions 72 hours into the future within a small variable, since no two storms are exactly alike, and so complex systems such as the atmosphere on Earth can have successful predictions made about them, despite whatever chaos theory may say to the contrary.

Supposition19: Global warming will, within 30 years, no longer be viewed as an imminent threat. Global temperatures stabilize within that time and begin to show little variation from year to year, with ocean levels well below the highs of thirty years previously.

Supposition20: The increased availability of water on land, and improved runoff from rivers to oceans begins to slow the extinction of species considerably in 20 years. The plant and animal life in the wild both on land and in oceans and fresh water habitats begins to show marked increases in populations in 30 years.

Supposition21: Desalinization plants will no longer be planned in twenty years, and in the case of those already existing, some are decommissioned and converted to sports stadiums within 20 years.

Supposition22: Within fifty years, an attempt to harvest electricity from lightning will have been made, and based upon actual data from the ongoing project, numerous small weather making stations also equipped to capture electricity will spring up around the world. By the end of the 21st century the number of such weather modification stations equipped to capture lightning and convert it into electrical current while producing adequate precipitation for the local area it occupies exceeds 100.

Supposition23: The encyclopedia of the world will begin to contain information about the type of weather modification discussed in this book within 20 years.

Don't jump, daddy! The internet has a way to save the farm!

Supposition 24: The Absolute Frame Of Reference will be worked out by astronomers within the next 50 years. Instead of finding our direction through empty space by accident, a computer program is developed that supposes it has the direction we are going, and compares that direction to a series of astronomical observations it has stored. This computer program compares billions ofpossible directions through space per hour and eventually finds which particular direction, among billions, matched astronomical readings most closely.

Supposition 25: Within 30 years the popularity and widespread use of the processes and theories described herein results in many countries easing religious restrictions and permitting many other things that were once considered unlawful.

Supposition 26: The internet will begin to contain this information within 10 years. The first places where the processes discussed here are found on the internet probably won't be in encyclopedia, but on other web sites, like "conspiracy theory" web sites, maybe some environmental and educational web sites.

These predictions all begin February 22, 2010. That was the date of release to the public of the first edition, and this second edition just continues the whole discussion a bit more. Even if the ideas here presented only reach a few people in the first few years after publication, eventually some interest should get generated, and two or three years shouldn't affect the predictions much.

Maybe a lot depends on the attitudes of people toward tampering with nature. The only way that attitudes can change and more and more people envision a brighter future involving a little custodial nature tampering is if books like this are written and read. I certainly got nowhere with the meteorological community by sending a one page description of this theory to various web sites.

CONCLUSION

This process in nature that I've tried to describe and talk up needs a full investigation by the scientific community. I watched the presentation about Ardi, the 4.4 million year old hominid fossil that was found in the highlands of Ethiopia, and all the nations and scientists who collaborated in the excavation and examination of the fossils, and it seems that a good deal of time and trouble was spent on it. Some time and trouble could as easily be expended in investigating the processes described in this book.

I referred to the same archeological find in the introduction of the book. That television presentation I watched broke a long standing paradigm. The mainstream archeological community had long discounted any hominid finds made that appeared to be from longer ago than 100,000 years, and that predisposition of most archeologists to ignore older finds became the subject of a book. "Forbidden Archeology"[11] by Michael A. Cremo and Richard L. Thompson explores a number of archeological discoveries "swept under the rug".

That book was a much more challenging bit of work than this one, and the authors did much better with it. My book is a mere magazine article in comparison to the wealth of information presented in Forbidden Archeology. Nonetheless, both sciences appear to have problems discerning the truth as it applies to its particular discipline. It seems to have paid off if mainstream television carries an archeology program about a much older hominid fossil than previously. The paradigm has shifted, in less than 15 years from the publication of Forbidden Archeology. This book doesn't have the wealth of research that Forbidden Archeology has, but you have to admit, cosmology, astronomy and astrophysics differ from archeology. One doesn't have specimens in sciences that deal mostly with theories and hypothetical questions.

Meteorology differs quite a bit from archeology as well. There are real things to study in meteorology, real possibilities uncatalogued that need more light shed on them. My hope is this book will help to serve as an impetus to move this idea further along, similarly to what happened in archeology. One would think that meteorology would get a more careful review, and more details about what has been discussed here about weather modification should eventually see a presentation on mainstream television.

The scientific community could expand on this issue to the point where my book here would look like a brief introduction. The resources and technicians the scientific community has could answer all the questions raised herein and give much more detail about these issues. Readers could help the author to see more development in this area just by being of that opinion. If enough people persist in wanting to see further investigation into something, it eventually happens. The thing every reader should want to know at this point is, what, indeed, will the scientific community make of all this eventually?

People like to know things. I do believe there is something to what I have tried to present. The invisible forces of electromagnetism and gravity are around us continually; that the two could combine to produce effects involving the water molecule, along with the hypothesized primordial specks such as described makes sense. It fits all the data. The seeing is conceptual in the case of primordial specks, and the forces of nature.

No other explanation or theory fits all the facts as well. As for how dangerous this kind of weather modifying activity could be, it doesn't get any more or less dangerous based on how many people on Earth know of it, it is an inescapable feature of having a tool capable of doing as described earlier. Knowing always involves an empowerment, and, surely, once enough people become aware of this and begin to discuss it, our birth as weather modifying beings will have begun, and before long, proper management of fresh water on a world wide basis will come easily and safely.

No need to do anything but get this confirmed by reputable scientists and placed in encyclopedia. Human self preservation will take it from there. We have a low tech tool with a wide range of uses, some of which were mentioned earlier. There is some work involved implementing the tool, some equipment, but surely possible almost anywhere on the globe. The questions each reader must ask themselves are: Does what Mother Nature does without human intervention have such divinity that they would prefer to suffer whatever that may be including all the bad types of weather that we have already discussed, and remain deluded, or would the reader rather not experience drought, flooding, tornados and hurricanes and be the wiser?

The question also remains whether weather events in the future will be done by nature alone, or with some assistance, and that question won't be easy to answer in a world where we don't learn this. That world will be a world where this feature, this process in nature will still exist, whether acknowledged or not. The beauty of nature stands a chance only if weather is optimized for the wild inhabitants. Traveling through Yellowstone National Park about 7 years ago we saw a Yellowstone with huge tracts of scorched burnt trees, from forest fires not long before. Would that have happened if man had been actively intervening with the proposed "new" technology when forest fires occurred? Was my appreciation of nature enhanced by the destruction caused by unrestrained mother nature? Of course, firefighters and equipment, planes and helicopters dumping water and chemicals were used to fight the fires, but is that the most effective way to combat a forest fire?

Weather disasters need to be reclassified. The possibility that these types of misfortune could be minimized, even eliminated completely, exists. No one is going to try to minimize a potential weather disaster if one appears likely without the knowledge of how to do so. Proof would, of course, be needed. One wouldn't base actions on controversial information, which is what this book amounts to prior to any scientific confirmation. Finding out late into the preparation of this book that the water molecule itself most likely factors quite heavily into what happens to the atmosphere, would indeed make it seem that the scientific community has not been exactly forthcoming with what it knows to be true.

Science and applied science are two different things.

One would have to group the process of placing metals strategically into the applied science category. That doesn't change things much, since encyclopedia already have allotted the term "weather modification" a place among all the subjects in the encyclopedia. There is already a place for a summary of the processes presented in this book.

Supposing that weather problems with flooding got out of control by 1930, could it really be possible that whoever was experimenting at the time decided that the whole process was too dangerous, and kept the matter to themselves? I think it likely something of the sort took place. A group of meteorologists, the Army Corps of Engineers, even some group like the Freemasons might have been doing the experimenting. Were it these last, secrecy is part of the agenda of that group. In that day and age it may have been possible to keep one of the more significant discoveries of modern times from public knowledge.

Today, however, numerous satellites continually view every inch of Earth from space. The internet connects every corner of the world to every other corner. The exact wrong thing to do in this day and age would be to try to continue to keep this a secret. Empowering everyone will eventually take away any plausible deniability when it comes to tampering with the weather. Any and all weather modification would be strictly controlled and monitored before long, once the initial idea gains acceptance.

Currently the weather is viewed as occurring naturally; I can't prove that some of the unusual weather that has occurred recently has been anything but that. Continuing to view the weather as occurring naturally in the future, in light of all that has been presented here, would be foolish in the extreme. It would be extremely helpful if one day some weather modification activity that got out of hand actually got found and removed by the authorities, preventing a weather disaster. If some event like that were to occur, and then be discussed in the news, it might start a change of attitude, a bit more optimism, about the whole idea.

The number of meteorologists, hydrologists, electrical engineers, and other professional people who should have this knowledge available to them is staggering. Bringing flooding under control in this day and age should be considerably easier than it was in 1930. Since that is apparently the only impediment to learning of this process, experiments with lead to raise air pressure should be explored as soon as possible. Negatively charged electromagnets could prove effective in repelling water molecules, as well. Finding a quantity of copper or lead or an electromagnet that is causing trouble, in this day and age, is going to be a lot easier than it was eighty years ago. If the governing body got truly serious about a zero tolerance policy toward weather tampering, SWAT teams could be descending from helicopters upon a location with falling barometric pressure in just a few hours. It would help if laws were passed first before drastic measures are taken. Discussion, experiment, and debate should get us closer to the truth in this matter. Nowadays the larger media organizations have become a power unto themselves. That power needs to turn its attention to this issue.

What agency makes the decision in a free society about the comparative safety of learning a process and remaining ignorant? My contention is that the public is the supreme ruler of the marketplace. The tiny minority that may do mischief are overwhelmed by all the professional people who are capable of putting these processes to productive and safe use. No group should be able to keep mankind in the dark like the last century.

With the dangers possible with the kind of activity discussed here, and the difficulties involved in getting a scientific discipline to change its stance regarding new discoveries, chances are the paradigm may not shift for a long time. Who knows when these new ideas reach an encyclopedia. That is up to the youth of the world, and upcoming generations. They will not be starting with as little as I did. One book, at least, points the way. No human wants to live like a horse, with blinders preventing them from seeing the big picture. Seeing the big picture is going to be difficult without reputable scientific help.

I'd like to include a conceivable excerpt from a hypothetical encyclopedia entry from the future…. Weather Modification; Any act or process created by human activity that has as its end result the changing of the weather. Recent discoveries suggest that weather systems can be created and destroyed by placing purified copper or lead in the hundreds of pounds in high locations. Copper tends to lower barometric pressure and produce storms, and lead seems to have the opposite effect. The storms brought on by copper in a high location usually take approximately 72 hours to develop give or take 8 hours. Places near an ocean to the west usually begin to experience precipitation 12 hours sooner than places further inland. Why these effects should happen isn't precisely certain, but much of the effects are probably due to the polar nature of the water molecule, and perhaps some primordial specks from the Big Bang also following the path of least resistance provided by the copper along with the water molecules. Lead seems to provide no particle with charge any clear path to follow,so the subsequent scattering of particles in the vicinity of lead in a high location raises air pressure, dispersing clouds. Electromagnets proved too unpredictable, but can also attract or repel water molecules, depending on the polarity. The authorities of most countries are quite adamant nowadays that no experiments of these types be conducted by anyone simply curious about the process. Meteorological communities worldwide have banded together to ensure that as few weather disasters as possible is the norm, so don't go trying this yourself! For further reading, see; Water Molecule, Weather Knowledge, Weather Laws, ….

Any rational being, having read what arguments have been put forth here, would conclude that it is our duty in the natural scheme of things to perform custodial work on the planet we live on. Had humans not arisen as the apex creature on this planet, some other animal would have eventually, and so, in the cosmic scheme of things, a creature at the top of the food chain will evolve, and eventually fill the position of the only creature on the planet capable of modifying the weather. Does the universe expect that creature to carry out his duty? No, the universe just makes it possible.

* References

1. Tesla: Man Out Of Time
   By Margaret Cheney   C.1981
2. Tesla: Man Out Of Time
   By Margaret Cheney   C.1981
3. Giant Molecules
   By Alexander Yu. Grosberg and
   Alexei R. Khokhlov   C.1997
4. Dark Cosmos; In Search Of Our Universe's
   Missing Mass And Energy
   By Dan Hooper   C.2006
5. Dark Cosmos; In Search Of Our Universe's
   Missing Mass And Energy
   By Dan Hooper   C.2006
6. Dark Cosmos; In Search Of Our Universe's
   Missing Mass And Energy
   By Dan Hooper   C.2006
7. Weather Modification By Cloud Seeding
   By Arnett S. Dennis   C.1980
   Weather And Climate Modification:
   Problems And Progress
   By Thomas F. Malone C.1980
   Weather Modification: Prospects
   And Problems
   By Georg Breuer   C.1980
8. Human Action   C. 1949
   By Ludwig Von Mises
9. Atlas Shrugged
   By Ayn Rand   C.1957
10. Modern Physics and Antiphysics
    By Adolph Baker   C.1970

11. Forbidden Archeology
   By Michael A. Cremo and
   Richard L. Thompson C.1993